EMPIRICISM, LOGIC, AND MATHEMATICS

VIENNA CIRCLE COLLECTION

VOLUME 13

EDITOR: BRIAN McGUINNESS

HANS HAHN

HANS HAHN

EMPIRICISM, LOGIC, AND MATHEMATICS

Philosophical Papers

Edited by

BRIAN McGUINNESS

with an introduction by

KARL MENGER

D. REIDEL PUBLISHING COMPANY

DORDRECHT : HOLLAND / BOSTON : U.S.A.

LONDON : ENGLAND

Library of Congress Cataloging in Publication Data

CIP

Hahn, Hans, 1879–1934.
Empiricism, logic, and mathematics.

(Vienna circle collection ; v. 13)
Bibliography: p.
Includes index.
1. Philosophy–Addresses, essays, lectures. 2. Mathematics–Philosophy–Addresses, essays, lectures. 3. Intuition–Addresses, essays, lectures. 4. Infinite–Addresses, essays, lectures. I. McGuinness, Brian. II. Title. III. Series.
B29.H27 193 80–11504
ISBN 90–277–1065–1
ISBN 90–277–1066–X pbk.

Articles I–VI translated from the German by Hans Kaal

Published by D. Reidel Publishing Company,
P.O. Box 17, 3300 AA Dordrecht, Holland.

Sold and distributed in the U.S.A. and Canada
by Kluwer Boston Inc., Lincoln Building,
160 Old Derby Street, Hingham, MA 02043, U.S.A.

In all other countries, sold and distributed
by Kluwer Academic Publishers Group,
P.O. Box 322, 3300 AH Dordrecht, Holland.

D. Reidel Publishing Company is a member of the Kluwer Group.

Printed in The Netherlands

TABLE OF CONTENTS

Introduction by Karl Menger ix

Editor's Note xix

I Superfluous Entities, or Occam's Razor (1930) 1
II The Significance of the Scientific World View, Especially for Mathematics and Physics (1930) 20
III Discussion about the Foundations of Mathematics (1930) 31
IV Empiricism, Mathematics, and Logic (1929) 39
V Reflections on Max Planck's *Positivismus und reale Aussenwelt* (?1931) 43
VI Review of Alfred Pringsheim, *Vorlesungen über Zahlen- und Funktionenlehre*, Vol. I, parts I and II, Leipzig and Berlin 1916 (1919) 51
VII The Crisis in Intuition (1933) 73
VIII Does the Infinite exist? (1934) 103

Bibliography of the Works of H. Hahn 132

Index 136

INTRODUCTION

The role Hans Hahn played in the Vienna Circle has not always been sufficiently appreciated. It was important in several ways.

In the first place, Hahn belonged to the trio of the original planners of the Circle. As students at the University of Vienna and throughout the first decade of this century, he and his friends, Philipp Frank and Otto Neurath, met more or less regularly to discuss philosophical questions. When Hahn accepted his first professorial position, at the University of Czernowitz in the northeast of the Austrian empire, and the paths of the three friends parted, they decided to continue such informal discussions at some future time – perhaps in a somewhat larger group and with the cooperation of a philosopher from the university. Various events delayed the execution of the project. Drafted into the Austrian army during the first world war, Hahn was wounded on the Italian front. Toward the end of the war he accepted an offer from the University of Bonn extended in recognition of his remarkable mathematical achievements.[1] He remained in Bonn until the spring of 1921 when he returned to Vienna and a chair of mathematics at his alma mater. There, in 1922, the Mach–Boltzmann professorship for the philosophy of the inductive sciences became vacant by the death of Adolf Stöhr; and Hahn saw a chance to realize his and his friends' old plan. It was mainly through Hahn's influence that the chair was offered to Moritz Schlick, then in Kiel. Soon after his arrival in Vienna, Schlick began to arrange discussions for a small invited group, with Hahn and Neurath as the first principal participants. Frank was in Prague where he had gone before the war as the successor of Einstein; however, he visited Vienna at least twice a year. Then there was Victor Kraft and, at Hahn's suggestion, the geometer, Kurt Reidemeister, who in 1923 came to Vienna for about two years. Soon after he left, Rudolf Carnap arrived and joined the group to which Schlick had also brought promising students of philosophy working in his seminars, among them Herbert Feigl and Friedrich Waismann. Thus, ultimately through Hahn's efforts, his, Neurath's and Frank's old plan had become a reality. "*Man kann*," Frank wrote in his obituary of Hahn in *Erkenntnis*, Vol. 4, "*Hahn als den eigentlichen Begründer des Wiener Kreises ansehen*." ("Hahn may be regarded as the real founder of the Vienna Circle.")

Secondly, it was Hahn who directed the interest of the Circle toward logic. Schlick, a student of Planck's and admirer of Einstein, had until then been mainly interested in the philosophy of nature and in epistemology up to (but not beyond) studies of the axiomatic method. Otto Neurath and Olga Hahn (the blind sister of Hans, later Otto's wife) had written – individually and jointly – papers on Boolean algebra[2] in the years 1909 and 1910; but thereafter and especially during the war Neurath's interests turned again to economics, sociology and history, while Frank was engrossed in the philosophy of physics and the study of causality. Carnap, having been a student of Frege's, was well-versed in logic, but was mainly concerned with the philosophy of science at the time he moved to Vienna. Hahn, however, right after his return, began an intense study of symbolic logic with an eye to related philosophical problems. In 1922 he offered a course on Boolean algebra.[3] During the year 1924/25 he conducted a memorable seminar on the *Principia Mathematica* of Whitehead–Russell in which, after some introductory lectures, he let advanced students, young Ph.D.s and lecturers report on the contents of the book, chapter by chapter.[4] This seminar had a very large audience and was of great influence not only on the development of many Viennese students of mathematics and philosophy but also on the trend of the discussions in the Circle.

Thirdly, until his untimely death in 1934, Hahn greatly contributed to the Circle as a prominent participant in the discussions. Since he was carrying out very interesting mathematical research[5] in addition to his extensive activity at the University he unfortunately found little time to publish many of the ideas he proposed in the meetings. But his penetrating criticism, the clarity of his ideas and his skill in presenting them greatly impressed everyone, and often influenced Neurath and Carnap as well as Schlick and Waismann. "One can say," Frank wrote in the obituary already quoted, "that in a certain sense Hahn was always a center of the group. He always represented its central ideas without entering into differences of opinion on side issues. No one knew as well as he how to present those leading ideas in such a simple as well as thorough way, in such a logical as well as suggestive form."

While Hahn's mathematical knowledge was unusually extensive,[6] his familiarity with traditional philosophy was more limited. His favorite author was Hume, whose works, as he once told me, he found not only intellectually delightful but also morally uplifting. Furthermore he greatly admired Leibniz to whose *identitas indiscernibilium* he gave much thought, and Bolzano, whose *Paradoxes of Infinity* he edited. Kant, on the other hand, he strongly disliked because of the changes – from chapter to chapter and often from sentence to sentence – of the meaning of the terms used in his writings. Of the more modern philosophers Hahn favored Ernst Mach; and during the early 1920's he developed a great admiration for the works of Bertrand Russell. He reviewed some of them in the *Monatshefte für Mathematik und Physik*. In one of these reviews Hahn suggested that one day Russell might well be regarded as the most important philosopher of his time – a statement remarkable at a period when few philosophers in Central Europe knew or even cared to know Russell's writings.[7]

The subsequent development of his views on Wittgenstein was described to me by Hahn himself upon my return to Vienna in the fall of 1927, when he and Schlick invited me to join the Circle. He asked me whether I had heard of the *Tractatus*. I said that some time ago I had started reading the book but had not continued beyond the first pages. "This was my original experience also," Hahn said, "and I did not have the impression that the book was to be taken seriously. Only after hearing Reidemeister give an excellent report about it in the Circle three years ago and then carefully reading the *entire* work myself did I realize that it probably represented the most important contribution to philosophy since the publication of Russell's basic writings. Yet we had controversies in the Circle about the book, and there were so many differences of opinion about details that, a year ago, Carnap suggested we should, in order to clear up the confusion, devote as many consecutive meetings of the Circle as necessary to a reading of the work paragraph by paragraph; and we have indeed devoted the entire past academic year (1926/27) to this task. To *me*," Hahn continued, "the *Tractatus* has explained the role of logic." In later writings, which are included in this volume, Hahn expounded his own (somewhat oversimplifying) view by describing logic as "a prescription for saying the same thing in various ways, and for extracting from what is said all that is (in a strict sense) connoted."

In the early 1930's, after Carnap had gone to Prague, a controversy about a related topic arose in the Circle when Waismann proclaimed that one could not speak about language. Hahn took strong exception to this view. Why should one not – if perhaps in a higher-level language – speak about language? To which Waismann replied essentially that this would not fit into the texture of Wittgenstein's latest ideas. The debate recurred several times up to Hahn's death: Hahn asking, "Why?" and Waismann answering, in the last analysis, "Because". Schlick sided with Waismann,

Neurath with Hahn. Kurt Gödel and I, though reticent in most debates in the Circle, strongly supported Hahn in this one.

Another topic that Hahn raised in the Circle was a view on history that he had developed – his contribution to the Neurath–Carnap program of Unity of Science. Despite great reservations that I had with regard to the general program I felt that Hahn's interesting and, as far as I knew, original idea was quite independent of those generalities. Its starting point was Poincaré's remark that physicists are not interested in historical propositions such as *Caesar crossed the Rubicon* because they yield no predictions: Caesar will not cross the Rubicon again. Hahn took issue with a part of Poincaré's remark. He pointed out that assertions about Caesar are based on countless old manuscripts, paintings, documents, excavations, coins and the like; and that the core of the proposition, which is capable of further verification or of falsification, actually is a *prediction*, namely, the prediction that all old manuscripts, paintings, documents etc. to be found in the future will be consistent with the present assertions.

On several occasions, the mathematicians in the Circle helped the philosophers by providing them with technical information – often concerning their own results. I remember, for instance, a discussion about inductive processes in physics. In order to obtain formulae describing actual as well as potential observations, Carnap and Schlick suggested interpolation – the passing of a polynomial through the finite number of observed data. Hahn convinced them of the impracticability of this idea by reporting results of a paper of his.[8] If one starts with a given continuous curve, C, the following phenomenon may occur: One begins with a finite set, S_1, of points on C and interpolates, thereby obtaining a curve, C_1; then one adds a few points to S_1 obtaining a set S_2 and by interpolation a curve C_2; and so on. However, the curves $C_1, C_2, \ldots$ need not approach C. For between the points of the set S_n, where C_n and C agree, the curve C_n may swing far away from C; and with increasing n, more and more oscillations with appreciable amplitudes may accumulate and prevent the curves C_n, C_{n+1}, . . . from approaching *any* continuous curve. *A fortiori*, the same may occur in the more realistic situation where no curve C is given at the outset, but only a (potentially infinite) set of points representing observed data.

Carnap and Neurath as well as Schlick and Waismann often gratefully acknowledged Hahn's role. So, during Schlick's absence on a visit to the United States in 1929, Frank asked Hahn to deliver the inaugural address at the First Congress of the Epistemology of Science (in Prague), while Neurath and Carnap asked him to be the principal signer of the pamphlet *Wissenschaftliche Weltauffassung*.

Hahn's address (Chapter II in this volume) formulated with masterful succinctness and precision a kind of creed for logical positivism and the scientific *Weltanschauung*. It contrasted them with mysticism, metaphysics and various other tendencies then prevailing in philosophy (especially in German philosophy), and also delimited them *vis-à-vis* pure empiricism and rationalism.

The pamphlet, whose genesis I had occasion to observe, was mainly written by Neurath[9]; Carnap cooperated to some extent; Hahn received the final draft.[10] It was well written and informative in various ways. Yet Schlick, to whom the booklet was dedicated, was not altogether pleased with it when he received it upon his return to Vienna. It lacked the depth and the precision of Hahn's address to the Prague congress. Also, it introduced political views and tendencies, which Neurath never completely separated from epistemological insights, whereas Schlick always insisted on the strict separation of the latter from value judgments of any kind. Moreover, some of those views were, if only in passing, presented as common to all members of the highly individualistic group, while some members, including Schlick himself, did not fully share them.[11]

In his later years, Hahn spent a minor but appreciable part of his free time on parapsychological studies. Many friends of his and admirers of his intellect found this interest of his very odd and wondered how the topic of parapsychology could even be broached in a group as strictly scientific in its orientation as the Vienna Circle. A partial explanation lies in the way Hahn's interest originated. In the first years after World War I a new influx of mediums had appeared in Vienna. They were viewed by the intelligentsia with the utmost skepticism. Finally, one day in the early 1920s, the newspapers claimed that two professors of physics at the University of Vienna, Stephan Meyer and Karl Przibram, had exposed the entire spiritualistic swindle. What had happened, the newspapers elaborated, was that Professor Meyer had invited many people to his house for a séance. The guests sitting in a circle and holding hands in a dark room, had clearly observed the phenomenon of levitation – more specifically, a figure in white rising several feet into the air. But at that moment the host unexpectedly turned on the lights and everyone could see that the apparition was none other than the tall professor Przibram who, in the dark, had managed to cover himself with a bed sheet and to climb on a chair. In the midst of general laughter the two physicists claimed that they had produced a levitation and exposed the mediums.

It goes without saying that the parapsychological groups were outraged; and for once, in a reversal of the ordinary situation, the mediums called all scientists swindlers. But there was also great indignation in the intellectual community; and a group including besides Schlick and Hahn the eminent neurophysiologist and physician Julius von Wagner-Jauregg, the physicist Hans Thirring and a number of others (most of them scientists) formed a committee for the serious investigation of mediums. Very soon, however, members began to drop out: first Wagner-Jauregg; soon after him Schlick. By 1927 apart from non-scientists only Hahn and Thirring were left in the group. They were, as they told me, not convinced that any of the phenomena produced by the mediums were genuine; but they were even less sure that all of them were not. They believed, rather, that some parapsychological claims might well be justified; and that certainly the matter warranted further serious investigation.

Hahn devoted a few interesting public lectures to the subject. I remember especially two points from these lectures. One was taken from the physiologist Charles Richet, an early French Nobel laureate, who suggested that one should imagine a world in which all men with the exception of a few scattered individuals had completely lost the sense of smell. Walking between two high stone walls one of those few might say, "There are roses behind these walls." And to everyone's amazement, his assertion would be verified. Upon beginning to open some drawer he might say, "There is lavender in this drawer," and if none should be found he would insist, "Then there *was* lavender in this drawer." And sure enough, it would be established that two years earlier there was indeed some lavender in that drawer. Are mediums with extraordinary perceptions and exceptional abilities in our world what the few people with a sense of smell are in Richet's?

The second point goes back to Hahn himself. Some mediums claim that great thinkers and poets speak through them while they are in trance, but actually those mediums utter only lines that are far below the level of their supposed authors. This well-known fact is usually construed as proving that the uneducated mediums simply say what *they* think their sources would say. Hahn, however, pointed out that many mediumistic revelations are *so* trivial and incoherent that even a medium with little education would not consider them as utterances of their supposed sources, and that in fact they are definitely below the *medium*'s own level. To Hahn this indicated that such chatter was not the product of the medium's conscious mind, but was generated subconsciously. Its very triviality combined with the tormented stammering in which the babble is frequently uttered, suggested to Hahn that in many cases one is dealing with a *genuine* phenomenon *of some kind*.

In his public lectures Hahn was of supreme clarity; but he also prepared his daily courses meticulously. He applied a technique that I have never seen anyone else carry to such extremes: he proceeded by almost imperceptible steps and at the end of each hour left his audience amazed at the mass of material covered. How stimulating his lectures were I have tried to describe – in Chapter 21 of my book *Selected Papers in Logic and Foundations, Didactics, Economics* (Vol. 10 of the Vienna Circle Collection) – by the example of the effect on myself of the first lecture he gave after his return to Vienna in 1921.

Politically, Hahn was a socialist of deep conviction and took part in several phases of the social-democratic movement. He never spared any pains to serve the public. In particular, he furthered adult education and did whatever he could for competent under-privileged students. Where he saw injustice or oppression, he tried to help the injured. Once on the street, when a coachman mal-treated his horse and Hahn's protest was ignored, he dragged the ruffian to the police. Hahn was respected even by his opponents.

KARL MENGER

NOTES

1 Hahn's first results were contributions to the classical calculus of variations. He then turned to the study of real functions and set functions, especially integrals. He further published a fundamental paper on non-Archimedean systems, and early recognized the significance of Fréchet's abstract spaces. In a paper introducing local connectedness he characterized the sets which a point can traverse in a continuous motion; that is, the continuous images of a time interval or a segment (now often called Peano continua). The paper is a classic of the early set-theoretical geometry. About Hahn's work after World War I, see note[5] below.

[2] Two of these papers, one of Olga's and one jointly written, are marked with an asterisk by C. L. Lewis in his book *A Survey of Symbolic Logic*, Berkeley, 1918, indicating that at that time Lewis ranked these studies among those "that are considered the most important contributions to symbolic logic."

[3] For administrative reasons, this lecture course was listed as a "seminar."

[4] I remember having reported about one chapter of the Principia myself before leaving Vienna in the spring of 1925. Contrary to what has been sometimes written about Hahn's seminars, Wittgenstein's name had, certainly up to that time, never been mentioned in them.

[5] After World War I, Hahn published volume I of his monumental *Theorie der Reellen Funktionen* (volume II came out posthumously). He then returned to the calculus of variations and to the theory of integrals from modern points of view. He applied some of his results to problems of interpolation, which later turned out to be of interest to the Circle (see Note[8] below). Perhaps most importantly, he developed the concept and parts of the theory of general normed linear spaces, simultaneously with and independently of Stefan Banach, after whom they are now called 'Banach spaces.'

[6] At one of the large yearly meetings of the German-speaking mathematicians a group including the most prominent members in Hahn's age bracket decided to test their general mathematical knowledge. According to a system agreed on in advance, they asked each other questions of the level and the type of those in doctoral examinations. Anyone missing an answer was excluded from further competition. One by one the contestents were eliminated leaving Hahn the ultimate winner, with the well-known analyst and number-theoretician Edmund Landau as runner-up.

[7] Apart from this remark, Hahn's reviews of Russell's, Meyerson's, and a few other philosophers' books are merely brief summaries of their contents, and have therefore not been included in this volume.

[8] *Mathematische Zeitschrift*, vol. 1. This is the paper referred to above in Note [5]

[9] It is only fair that a translation of the pamphlet has been included in Neurath's collected papers, *Empiricism and Sociology*, (Volume 1 of the Vienna Circle Collection, Chapter IX).

[10] Hahn signed even though he would have written the pamphlet somewhat differently and was not in complete agreement with all details – one of the concessions he was occasionally prepared to make for the sake of peace. He had an irrepressible penchant for mediating between conflicting views and between quarreling people.

[11] For the same reasons and because of what I regarded as superficial views on social sciences I felt myself somewhat estranged by the pamphlet – in fact to the point where I asked Neurath to list me henceforth only among those *close* to the Circle (Cf. *Erkenntnis*, 1, p. 312). And the pamphlet alienated Gödel even more.

EDITOR'S NOTE

This volume brings together for the first time the small but influential corpus of philosophical writings of a celebrated mathematician. Without them a collection representing the Vienna Circle would be incomplete, for he was one of its founders and it owed much to his inspiration. We omit only *Logik, Mathematik und Naturerkennen* (1933), reserved for a separate volume devoted to the 'Einheitswissenschaft' series. One item included (Chapter V) has not been published before – notes, evidently, for a lecture in answer to the criticisms of positivism issued by Max Planck in a famous lecture of 1930. Professor Karl Menger found the manuscript in a copy of Planck's lecture in his possession and generously put it at our disposal.

For permission to print the items included we thank Frau Nora Minor, Hahn's daughter; Felix Meiner Verlag Hamburg in respect of Chapters II and III; the Academy of Sciences of the German Democratic Republic in respect of Chapter IV; the *Göttingsche Gelehrte Anzeigen* in respect of Chapter VI; and Franz Deuticke of Vienna in respect of Chapters VII and VIII. For these two last chapters, by kind permission of Simon and Schuster Company, we made use of an anonymous translation published in James R. Newman's *The World of Mathematics* (New York, 1960ff.). We were unable to trace the successors of A. Wolf of Vienna (the original publishers of Chapter I).

B. McG.

SUPERFLUOUS ENTITIES, OR OCCAM'S RAZOR *

It seems to me that in the bewildering variety of philosophical systems it is possible to distinguish two main types: those systems of philosophy that *affirm* the world and those that *deny* it. The world-*affirming* philosophy builds wholly upon the world revealed to us by the senses; it takes this world as it presents itself: in its inconstancy, its irregularity, its motley, and tries to find its way about it, to come to terms with it, and to enjoy it. It takes as real only what is revealed by the senses; it finds it abhorrent to seek out other kinds of entities outside the sensible world. On the other hand, the world-*denying* philosophy distrusts the senses, takes the sensible world for sham and fraud, for mere *appearance*, and searches for true entities, for true *being*, behind the world of appearance conjured up by the senses. The ultimate cause for such different kinds of philosophizing would seem to be certain basic psychological attitudes: a certain optimism on the one hand and a certain pessimism on the other. Some one who feels at home in the sensible world, who is capable of deriving pleasure from it, will not look out for entities behind it; but some one who does not find his satisfaction in it, who has been denied the enjoyment of the world of the senses, will seek refuge in a world of different kinds of entities. And so the world-denying philosophy proves to be a means, employed over and over again, of consoling the masses of those who had reason to be somewhat dissatisfied with *this* world with the prospect of another world; and it is perhaps owing mainly to this circumstance that for two millennia the

* Published as a pamphlet in 1930 by A. Wolf of Vienna.

world-denying philosophy has been able to enjoy a virtually unchallenged rule, a rule that is being seriously undermined only in our own time. But this – as I should like to call it – political and economic role of the world-denying philosophy will not be discussed here. We must, however, take a look – if only very briefly and schematically – at the lines of thought that have led away from the sensible world.

We are first referred to cases where we are apparently deceived by the senses. For example, a stick placed in water seems to us to be bent, but in reality it is surely straight; a mirage shows us a lovely palm grove in the middle of the desert, but when we get to the place where there is supposedly an oasis, we find nothing there but sand; we see a rainbow in vivid colours at a particular place, but when we get to that place, there is only rain and nothing else. And now we are told: whoever lies once is not to be believed. Since our senses *sometimes* deceive us, perhaps they *always* deceive us. If we look through a red glass, everything looks red; if a demon always held a red disc before everyone's eyes, should we not believe that everything *is* red? Perhaps there really is such a demon constantly playing tricks on us because he enjoys making fools of us? Or perhaps it is not a demon but an all-good and all-just God who is punishing us for having once done something which annoyed him?

It may have been lines of thought such as these that led to distrust of the sensible world. But if we distrust the senses, what, then, should we trust? And here we are told: Trust *thought*. Unlike the senses, which apprehend mere appearance, thought, we are told, apprehends true, substantive being, and this is why true entities must be like the concepts in our minds. But these concepts appear to be quite different from the objects of the sensible world. In the sensible world we find many horses, but there is only *one* concept 'horse'; a horse of the sensible world is born; first it is young, then it grows old, and then it dies; but the concept 'horse' is not born, does not grow older, and does not die; the individual horses of the sensible world move and change, come into being and pass away, but the concept 'horse' is unchangeable and immovable, and is not subject to becoming or to passing away. And so Plato came to the conclusion that the world of our senses – the motley, protean, ever-changing world where everything comes into being and passes away and where nothing endures – is not the world of true being; the world of true being is a world of *ideas*, of which our concepts furnish us a likeness; and in it is enthroned, somewhere and somehow, the idea 'horse', ungenerated and incorruptible, immovable and unchangeable and uniform, whereas the horses of the sensible world have being only to the extent that they participate in the idea 'horse' – though what all this is supposed to mean is hard to say.

The philosophers of the Eleatic school, who were roughly Plato's contemporaries, advanced an even more abstruse doctrine. While Plato assumed only *one* idea 'horse' in the world of true being in contrast to the many horses of the sensible world, he nevertheless taught that there were many different ideas in the world of true being: an idea of man and of iron, of the true and the good and the beautiful, an idea of two, of three, etc. But to the Eleatics all multiplicity seemed incompatible with true being, and they taught that true being is one, ungenerated, incorruptible, unmoved, unchangeable, and undifferentiated.

The world-denying philosophy thus searches behind the flux of appearances it distrusts for an immovable pole it can trust. But wherever we hear the word 'pole' outside mathematics and geography, extreme caution is in order. Words like 'pole', 'polar', 'cosmic', among many others, belong in the rogues' gallery of philosophical terms, and while the propositions in which these words occur sound very profound, most of them are in reality entirely without foundation.

It was in Plato's time, four hundred years before the birth of

Christ, that the world-denying philosophy prevailed over the world-affirming philosophy of Democritus, and – though it was somewhat mitigated by Aristotle – it remained dominant from then on throughout antiquity – in spite of strong counter-currents like those of the Epicureans – throughout the middle ages in the form of the scholastic philosophy of the Church, throughout the early modern period in the form of rationalism, and into our own day in the systems of German idealism – and how could it be otherwise? The Germans are known after all as the nation of thinkers and poets. But gradually a new day is dawning, and the liberation comes from the same land that gave birth to political liberation, viz. England: the English are known after all as the nation of shopkeepers. The most illustrious names along the road to liberation are John Locke, David Hume, and our contemporary Bertrand Russell, the English nobleman who served time in prison during the war for his anti-militarist attitude. And it is certainly no accident that the same nation gave the world democracy and the rebirth of the world-affirming philosophy, and no accident that the land that saw the beheading of a king also witnessed the execution of metaphysics. For all those other-worldly entities of metaphysics – Plato's ideas, the Eleatics' One, Aristotle's pure form and first mover, the gods and demons of the religions, and the kings and princes of the earth – all share a common fate, and when the emperor falls the duke must follow suit.

Yet the weapons of the world-affirming philosophy are not the sword and the axe of the executioner – it is not as bloodthirsty as all that – though its weapons are sharp enough. And I want to talk today about one of these weapons, viz. *Occam's razor*. William of Occam, the *doctor invincibilis* (invincible doctor) of scholastic philosophy and the *venerabilis inceptor* (venerable founder) of the nominalist school was – what a remarkable coincidence! – an Englishman – like John Locke, like David Hume, like Bertrand Russell. He lived from 1280 to 1340, in the heart of the middle ages. And his razor, which sweeps away other-worldly entities as a scythe sweeps away the grass in the meadow, is the proposition *Entia non sunt multiplicanda praeter necessitatem*. In translation: One must not assume more entities than are absolutely necessary. The word *entia* ('being') would be underlined in red if used by a high-school student in his homework; for classical Latin, known as the most logical of all languages, did not have a present participle for *esse* ('to be'); it could not say 'being ; and Seneca complains that when one wants to translate the word ὄν ('being') of the Greek philosophers, one has to say *quod est* ('that which is'); the word *ens* is a late artificial formation.

Let us return to Occam's guiding principle. To understand the sense of this proposition within its historical context, we must say a word about the mediaeval controversy about what are called universals. Universals are general concepts: the concept 'horse' as opposed to the individual horses which are really there, which are to be found in the world. The philosophy that denies the sensible world, in continuing Plato's theory of ideas, asserted that corresponding to these general concepts there were entities having a real existence in some world of ideas; its representatives are therefore called *realists*. On the other hand, the philosophy that affirms the sensible world says that it is enough if there are horses to be found in the sensible world; it is superfluous to assume that there exists an idea 'horse' besides them, an entity corresponding to the concept horse. *Sufficiunt singularia, et ita tales res universales omnino frustra ponuntur*. Or, in translation: "Individuals suffice, and so it is entirely superfluous to assume those universals." And since they are superfluous – and this is the application of Occam's razor – away with them! The world-affirming philosophy thus took the view that there are no entities corresponding to the universals; rather, these universals are mere names, *nomina*, and this is why the adherents of this line of thought are called *nominalists*. To conclude, the realists were world-denying, the nominalists world-affirming, and Occam was the leader of the nominalists.

In this connection we must also say a few words about Occam's life. Occam was a Franciscan – at that time religious orders were the bearers of intellectual life – and he was not an unbeliever; that would be too much to ask of a philosopher of the time. But he was nevertheless a free-thinker, and his free-thinking came out in his saying that thought could never yield the dogmas of the church or the existence of God. We shall see presently how closely this doctrine is related to the opposition between world-denying and world-affirming philosophy. I once read in a French dictionary of philosophy: *Téméraire en philosophie, il fut insubordonné en religion*: "Being audacious in philosophy, he was insubordinate in religion." He was expelled from the Franciscan order, banished from the universities, and summoned to the papal court at Avignon; he was even sentenced to life imprisonment, though he did not serve out his sentence. The Popes at that time claimed that the spiritual power of the Pope was pre-eminent over the temporal power of emperors and kings. Occam joined Ludwig of Bavaria, the German emperor and opponent of the pope. Thus he took up the call of the Ghibellines. And the struggle between the Ghibellines, who favoured the temporal power of the Emperor, and the Guelphs, who favoured the spiritual power of the Pope, this centuries-long struggle between Ghibellines and Guelphs is only part of the eternal struggle between world-affirmers and world-deniers, as is the struggle between nominalists and realists in philosophy.

Let us continue with the opposition between world-affirming and world-denying philosophy. We have seen that the former trusts the senses and looks for true entities in the sensible world, whereas the latter distrusts the senses and relies on thought, which, in its opinion, provides access to the world of true entities. Our conviction is that the latter greatly overestimates thought. For all thought is nothing but transformation; thought can never lead to anything new. It is a fundamental truth, discovered very recently, that all logical thought is tautological; it can only help us to say in another way what has already been said, and it can never help us to say anything new. "All men are mortal, Alexander is a man, therefore Alexander is mortal" – this example shows roughly what all thought looks like. But if Alexander is a man and I say "All men are mortal", then in saying this I have already said that Alexander is mortal; hence the conclusion "Therefore Alexander is mortal" does not furnish anything new. Thought is therefore perfectly incapable of leading us from the sensible world into a world of different kinds of entities; there can be no logical chain of thought beginning in the sensible world and ending in a world of different kinds of entities. And now we see again how close Occam is to us in spirit. As we teach today that thought cannot lead out of the sensible world to other-worldly entities, so he taught at the time that thought cannot lead to God or to the dogmas of the church; these are left to a different domain, that of faith. While he believed in them, many nowadays do not believe in them; but *this* difference is, epistemologically speaking, not a big one, just as it makes no difference, epistemologically speaking, that one person loves narcotics and another one does not.

If the first basic error of the world-denying philosophy is to overestimate thought, its second basic error is to overestimate *language*. The propositions of our language are essentially built in such a way that they ascribe a predicate to a subject, or alternatively, several subjects. 'This piece of chalk is white." "This piece of chalk and that piece of chalk are white." So far everything is in order. But language also says: "This piece of chalk and that piece of chalk are the same colour." Thus language acts as if the property "same colour" was being ascribed to each of the two pieces of chalk – as the property "white" was being ascribed to each of them previously. But this is patent nonsense: "same colour" is not a property of an individual, unlike "white" for example, but a relation holding between two individuals. The world does not have a simple subject–predicate structure, as language makes it appear. On the contrary, it seems that, as our knowledge increases, subject–predicate structures get pushed further and further back in favour of relational structures. Einstein's theory of relativity is a giant step forward in this direction. In propositions like "It is raining", "It is snowing", "It is thundering" the subject–prédicate form is carried to an extreme.

In any case, it is a big mistake to infer the structure of the world from the structure of language. While there is much talk about the profundity of language, this is for the most part rhetoric. Language originated in primaeval times, in times when our ancestors were on anything but a profound level. Our cousins, the anthropoid apes, are certainly not very profound thinkers, and it is absurd to suppose that the evolution of mankind from an apelike state to its present one passed through a period of particularly profound thought. Language is accordingly an extraordinarily imperfect instrument, which constantly reveals the primitive grimaces of our primaeval ancestors, as when a highly enlightened free-thinker cannot help feeling uneasy if he is the thirteenth person at the table, or if he meets an old woman on his way to an important undertaking.

Just as the world-denying philosophy overestimates the power of thought, so it overestimates the significance of linguistic forms. It may well be said of it that where a subject is present, an entity is bound to appear in due course. And it is almost surprising that there is no chapter in the history of philosophy where the "it's" of the propositions "It is raining", "It is thundering", and "It is snowing" are hypostatized as entities and the most acute and profound thought is devoted to the question whether the raining *it*, the thundering *it*, and the snowing *it* are the same entity or different entities and how they are related to one another.

As an example of the tendency to overestimate language, let us consider very briefly the doctrine of what are called *impossible objects*. "Wooden iron does not exist." This, it used to be said, is a true proposition; it correctly ascribes a predicate to its subject, "wooden iron", namely non-existence. And this led to the further conclusion that since something can be correctly ascribed to wooden iron, it must somehow be something. In spite of its non-existence, there must therefore be an object, an impossible object, wooden iron. This is what the doctrine of impossible objects maintains; and it even claims to have significance for mathematics, arguing that mathematics operates meaningfully with impossible objects in its indirect proofs. This doctrine rests entirely on the misleading character of linguistic form. But today we have a language which is much less imperfect than the various word languages: this is the language of *symbolic logic*. Expressed in this language, and retranslated into ours, the state of affairs in question looks like this: The statement "x is iron and x is wooden" is false for any object x. Having corrected language in this way, we see that no impossible object occurs in the new formulation. If I run through all real objects x in the real world, I find that none of them is both iron and wooden. I need not therefore assume an impossible object in order to make sense of the proposition "Wooden iron does not exist"; and now Occam's razor comes into play: impossible objects are superfluous entities; hence, away with them! We do not need a sphere where impossible objects have a kind of being or a kind of subsistence, or where they lead a shadowy life.

As a further example of the operation of Occam's razor let us consider the doctrine of *time and space*. Let us speak first of time. Metaphysically-inclined philosophers assume an entity, "time", in which the events of our sensible experience flow as in a river. This entity time, they tell us, is composed of simple entities, viz. extensionless moments of time or points of time. Let us look at this from the standpoint of the world-affirming philosophy. We then have the processes we experience in the sensible world and nothing else. Some of these processes we experience as (wholly or partly) simultaneous. In the case of some others, we experience one as earlier and another one as later. And these experiences of simultaneity and succession are just as real and just as primordial experiences as our visual or our tactile experiences for example. We must not let ourselves be confused by language at this point, by the words, "at the same time": it is not as if I first had to have knowledge of time and of moments of time before I could find out that two events occurred at the same time or at the same moment of time. Simultaneity is primordial; we experience it as we experience colours and sounds; but no man has ever experienced a moment of time and no one ever will.

We also must not let ourselves be confused by Kant's world-denying philosophy, according to which simultaneity and temporal succession are subjective ingredients added by man to a timeless world of "things in themselves". Simultaneity and the relations of earlier and later between experiences are real, like red and green, like hunger and love; but no man has ever experienced anything like a "thing in itself" and no one ever will.

Let us now imagine a group of processes, of experiences, that are all wholly or partly simultaneous, which means that they overlap in time (wholly or partly). During a certain time-span (to use at first the ordinary form of expression) all of these experiences will therefore be going on together – some will already have begun earlier, some will still go on later. Let us now consider in addition still further experiences which are all (wholly or partly) simultaneous with all the experiences just considered: the time-span during which these experiences are going on – i.e., all the ones we considered previously and the additional ones we are now considering – will have become shorter. Let us now consider a class of experiences occurring partly simultaneously such that it can no longer be extended, i.e., a class of experiences of which any arbitrary number occurs (partly) simultaneously but which is so comprehensive that there is no further experience which would be (partly) simultaneous with all experiences of this class: the time during which they occur altogether reduces now to a single moment of time. Up to now we have used the customary form of expression, but now let us say with Russell: A moment of time *is* nothing but such a class of simultaneous experiences, one that can no longer be extended.

What has been said is, of course, a mere hint. But we must go briefly into two plausible objections. The first objection is this: it could surely happen that all of the experiences of such a class of simultaneous experiences, one that can no longer be extended, were to go on together not only at a single moment of time but for a whole span of time. This objection rests entirely on the prior assumption, which is in the spirit of the world-denying philosophy, that time is an independent entity. Our reply to this is as follows: if all experiences of the class of experiences just mentioned were really to go on together during a whole span of time, this would mean that, in this whole span of time, none of my experience had a beginning and none of them an end; I would therefore experience no change at all in this whole span of time; and thus I would have no means of finding out that a real span of time had elapsed and not just a single moment. But that which cannot be found out in any way whatever also has no sense whatever.

The second objection is this: a moment of time is surely something quite different from a class of experiences! But this objection too has no sense whatever. If we knew what a moment of time was, if we could experience a moment of time, then we could compare the experience "moment of time" with the class of experiences considered above, and we might perhaps be justified in saying that it was something different – as we know the colour red by experiencing it, and if some one wanted to say, "The colour red is a pile of bananas", we should be justified in replying: "That is not true". But since we *cannot* experience a moment of time, the present objection loses all sense.

This Russellian definition of a moment of time mentions only experiences; it remains world-affirming throughout; it makes no claim to any other-worldly, extrasensory entities. And it can be shown that these Russellian moments of time do all the work that physics demands of moments of time. It is therefore superfluous to assume that besides experiences there exist moments of time in which these experiences take place, and superfluous to assume that there exists a time whose extensionless parts would be these moments of time. We need nothing other than our experiences with their relations of simultaneous or earlier and later which we can themselves experience. And since it is superfluous to assume that time itself and moments of time themselves exist, we again apply Occam's razor and say: Away with them, and away with all philosophical pseudo-problems attached to the metaphysical existence of time!

Now something like what has been said about time can also be said about space. It is usually assumed that there is an entity, "space", composed of indivisible, unextended entities, viz. points; this space is the reservoir in which the processes take place of which our senses give us testimony; and in this space, existing in itself, the bodies of the sensible world float about like fish in an aquarium. As a child I was told that it was very difficult to grasp what such an extensionless point was, and that it was very difficult to imagine such a thing. Very difficult? This is an extraordinarily mild way of putting it. It is not difficult, it is completely impossible. Just as no man has ever experienced anything like a moment of time, so no one has ever experienced anything like a point of space.

It was thought till recently that geometry dealt with space and spatial points, and laymen probably thought that, even if they themselves did not succeed in grasping what a point of space was, mathematicians at least had a perfect grasp of it. But this was an enormous error: mathematicians had no better grasp of it, and they were no more capable of saying what a point really was than the nearest layman. And since it seemed to them quite hopeless that they would ever find out, whereas – apart from this blemish, that they could not quite say what a point was – the science of points, geometry, worked perfectly smoothly, mathematicians took an amazingly simple way out, saying that the question of what a point is does not belong to mathematics, that it is none of their business, and that one can pursue the science of points perfectly well without knowing what a point is. This does at first sound rather peculiar, but it is very far from being nonsensical, and it even led to very considerable advances in our knowledge of the structure of the science: it gave rise to the so-called axiomatic method with its theory of implicit definitions and hypothetico-deductive systems. But we cannot go into these things here; all that matters to us is the following: when mathematicians say, "A problem is none of our business", this does not solve the problem, nor does it get rid of it. There remains therefore the troublesome question, "What is a point?" And here we proceed just as we did in our inquiry into moments of time. While we find no points in the sensible world, we do find bodies, and we know what it means to say that two bodies have a piece in common, or that they overlap wholly or partly. We will at least assume that we know this, even though there still lurk difficult problems here, though problems of a somewhat different kind from the one to be solved now. Next, we consider again a class of bodies all of which have a piece in common. The more such bodies we consider, the smaller the piece becomes that is common to them all. We consider again a class of such bodies such that it can no longer be extended and say with Russell; Such a class of bodies *is* a point. We are now, of course, poles apart from the former definition: "A point is an extensionless part of space", which failed to express a thought. We are no longer dealing with an entity, "space", composed of parts – and in particular extensionless parts – that must be taken into account; we are dealing only with the bodies furnished to us by the sensible world; and certain classes of such bodies we call points. We do not therefore have to assume an entity essentially different from the sensible world: a space existing in itself with bodies swimming around in it and points existing in themselves of which this space is composed. We do not need them; and so Occam's razor comes again into play: "Away with them!" All we need is the bodies of the sensible world and we can understand everything that geometry needs.

And since we are already on the subject of mathematics, let us also speak of *numbers*. Metaphysically-inclined philosophers conceived of numbers as entities having some kind of mystical existence. According to Plato, there is in the world of ideas an idea of two, an idea of three, etc., and some metaphysicians believed that these numerical entities had secret miraculous powers. The number seven was favoured above the rest; it was supposed to be especially sacred: there are seven sacraments, though also – if I am not mistaken – seven cardinal sins, but this is supposedly a matter of polarity. And at one time the entity seven did in fact perform an act of creation: it produced the spectral colour indigo; for without indigo there would only be six spectral colours, and it would be a very deplorable defect in the world if there were not just as many spectral colours as there are holy sacraments; and this is why children learn about the colour indigo in optics, even though they may never come across it anywhere else.

The case of numbers is similar to that of time and space. As we have seen, we experience simultaneous events, but we never experience time itself or a moment of time. Likewise we find in our experience sets consisting of three or of five objects, but we never experience the number three or the number five. Now what do we really mean when we say that a set of objects consists of five objects? What we mean by it is that we can count them off with the help of the fingers of one hand; there are as many objects as one hand has fingers, or more precisely: the objects of this set can be correlated with the fingers of one hand in such a way that each object corresponds to one and only one finger and each finger to one and only one object. The sets consisting of five objects are therefore those and only those sets whose objects can be correlated in the way described with the fingers of one hand, and they thereby differ from all other sets containing more or less than five objects. And now we can say – and this is the very same idea that led us to the definition of moments of time and points of space: the number five *is* nothing but the class of all sets whose members can be correlated in the way described with the fingers of one hand. Here too it should not be objected that, for God's sake, a number is surely something different from a class of sets. Since we do not experience numbers as we experience colours and sounds, numbers are nothing in and by themselves; they are not given to us as colours and sounds are given to us; they must be constituted out of the given, i.e., out of the sensible world, and constituted in such a way that they do all the work that numbers are supposed to do both in daily life and in science. Following Frege and Russell, we have indicated how numbers can be constructed out of the given sensible world in such a way. Although I cannot, of course, show this here in detail, they do, in fact, do all the work that numbers in daily life and in science are supposed to do, and this is all we need. We manage perfectly well with the objects of the sensible world, and we never find it necessary to assume numbers as extra entities of their own with their own kind of existence besides or behind the sensible world. And since we do not need numbers as entities of their own, since we manage without such entities – and this is again an application of Occam's razor – neither do we want to assume such entities. There are only the objects we count, but there are no numbers as extra mystical entities of their own.

There are plenty of other examples where Occam's razor frees us from more or less shadowy, more or less obtrusive, superfluous entities. I will say a few more words about some especially obtrusive entities. I have spoken already of the subject–predicate structure of language and its disastrous effect on the world-denying philosophy when it inferred the structure of the world from the structure of language. We say "This piece of paper is now smooth, now folded up, now crumpled into a ball", and we find in this proposition a persisting subject (the piece of paper) and a changing predicate (now smooth, now folded up, now crumpled into a ball). Philosophers misled by language inferred from this that the world was built the same way, that corresponding to the "persisting" subjects of language there existed in the world persisting substances which – corresponding to the changing predicates – bore changing properties (qualities). But where do we experience a persisting substance? Nowhere! I do not experience a piece of chalk as a persisting substance or as a bearer of its qualities. What I experience is its white colour, the feeling of resistance or hardness when I reach for it, and its shape, which I can see as well as touch. But the shape I see changes when I put the piece of chalk elsewhere or when I myself step elsewhere. I do not experience that what I saw earlier and what I see now is *the same*. In short, we do not experience the familiar objects in our environment as persisting substantial entities; what I experience is changing colours, shapes, degrees of hardness, etc.

Have we now perhaps reached a point where we have to assume – besides that which we experience: the colours, sounds, etc. – still other entities which we cannot experience: the persisting substances which bear the changing colours, shapes, degrees of hardness we experience as their qualities? Here too we are being helped by the same idea that allowed us to conceive space, time, and number without having to go beyond possible experience, without having to assume entities of their own besides or behind those of our sensible world. What we call "this piece of chalk" *is* nothing but the class of colours, shapes, degrees of hardness, etc. of which it is supposedly the bearer. If we follow up this idea, which we owe to the world-affirming English philosophers and which was pursued with special emphasis by Ernst Mach, the great Viennese physicist and philosopher whose name our society bears, we learn that it is never necessary to assume behind or beneath the entities of the sensible world we can experience still other entities we cannot experience, viz. substances; and now even substances are swept away by Occam's razor with a single stroke: their much-praised persistence does not withstand the sharpness of this razor. But I need not go further into this, since this question will certainly be discussed in depth in a later lecture to our society.

But we still have to come to grips with a final deep and difficult problem. Let us look back. We have seen that moments of time are certain classes of experiences, points of space certain classes of bodies, and numbers classes of sets (and sets themselves are classes of objects); we have also seen that the objects in our environment are classes of colours, shapes, degrees of hardness, etc. Again and again we have thus been led back to *classes*. It therefore seems that we cannot after all manage with the entities of the sensible world alone, but that we must assume, besides these entities of the first order, still other entities of a higher order which will withstand Occam's razor, namely classes of entities of the first order, further, classes of such classes (numbers were such classes of classes), and perhaps still higher orders. I can only mention here that at this point too the world-affirming philosophy has shown us the right way: Russell has shown how every meaningful statement in which the word 'class' occurs can be transformed into a statement which mentions only objects themselves and which no longer mentions classes of objects. To give only one obvious example of this, by the proposition "The class of lions is contained in the class of mammals" we surely mean nothing more than "every lion is a mammal", i.e., to make it somewhat more explicit, every object having the defining properties of the concept "lion" has also the defining properties of the concept "mammal", and this formulation mentions only objects of the sensible world which we can experience, and it no longer mentions "classes". In the final analysis, classes too prove to be superfluous entities, and there remains nothing over and above the entities of our sensible world which we can experience.

We have now seen Occam's razor at work, and we may say that it works well. The world-denying metaphysical philosophy has beaten up enough lather to make shaving easy. And what remains is a world cleansed of all shadowy half-beings which were supposed to dwell somewhere between being and non-being – such as impossible objects, universals, empty space, and empty time; cleansed also of all presumptuous entities claiming a stronger and more persistent mode of being than the motley, protean entities of our sensible world – such as ideas and substances or anything else they may be called.

But, it will be said, this world-affirming philosophy which contents itself with the sensible world and asks for nothing higher or deeper is – true to its origins – a veritable shopkeepers' philosophy, a philosophy for boors and philistines, a philosophy for non-poets unendowed with imagination; no world is beautiful save one in which fleeting phenomena rest on massively persisting substances below and are governed from above by ideas – in a highly spiritualized and sublime form – as Goethe depicts them:

> Goddesses reign in majestic solitude,
> No space surrounds them, much less time,
> To speak of them – is an embarrassment!

The latter we are only too ready to admit.

It may be that the world was very beautiful when a god or demon could be scented on every path; it may be that it was a very poetic time when all sorts of higher beings had to be constantly kept in a reasonably good mood by prayer and sacrifice, and all sorts of evil spirits had to be kept in check by magic spells. And even today those who are convinced that they can banish misfortune by spitting three times may be very happy, and I can well appreciate the elation of a man who, having broken a mirror, now lulls himself into the conviction that he will not get married for seven years. But we are still of the opinion that all this is superstition, and we must still take the view that it is right not to believe in these things, no matter how pretty or poetic they may be. And so it is no argument against Occam's razor to say what a great pity it is to sweep away all those superfluous entities because they were so beautiful and poetic.

But – and now we are returning to our point of departure – people have different tastes. One person enjoys multiplicity, the variegation and capriciousness of change, the confusion in the sensible world which is so difficult to unravel, and he therefore continues to affirm this world, whereas another who has been denied the pleasures of the sensible world denies this world and tries to make up other worlds behind it. But we, the adherents of the world-affirming philosophy, regard this as a side-track, a blind alley, a disease from which mankind has suffered for millennia, and it is our intention to free mankind from this nightmare. And if a man emerges from the mystical gloom of the world-denying philosophy and embraces the simple, transparently clear teaching of the world-affirming philosophy, we greet him with these words:

> Your emerge from death, uncertain sufferings,
> All that was faint and mean and vague,
> And learn with unobstructed vision to distinguish
> The sickly dawn from the clear day!

THE SIGNIFICANCE OF THE SCIENTIFIC WORLD VIEW, ESPECIALLY FOR MATHEMATICS AND PHYSICS*

The name 'scientific world view' is intended both as a confession of faith and as a delimitation of a subject:

It is to *confess our faith* in the methods of the exact sciences, especially mathematics and physics, faith in careful logical inference (as opposed to bold flights of ideas, mystical intuition, and emotive comprehension), faith in the patient observation of phenomena, isolated as much as possible, no matter how negligible and insignificant they may appear in themselves (as opposed to the poetic, imaginative attempt to grasp wholes and complexes, as significant and as all-encompassing as possible).

And it is to *delimit our subject* from philosophy in the usual sense: as a theory about the world claiming to stand next to the special sciences as their equal or even above them as their superior. For in our opinion, anything that can be said sensibly at all is a proposition of science, and doing philosophy only means examining critically the propositions of the sciences to see if they are not pseudo-propositions, whether they really have the clarity and significance ascribed to them by the practitioners of the science in question; and it means, further, exposing as pseudo-propositions those propositions that pretend to a different, higher significance than the propositions of the special sciences.

We confess to carrying on the *empiricist* tradition in philosophy, and as such stand firmly opposed to all *rationalism*, whether it regards thinking as the sole source of knowledge or as a source superior to experience. But we also stand opposed to the *dualist*

* First published in *Erkenntnis* **1** (1930–1).

tradition, which wants to look at thinking and experience as two independent and co-equal sources of knowledge. We believe, rather, that experience or observation is the only means of knowing the facts that make up the world, and that all thought is nothing but tautological transformation. This will have to be explicated more fully; but let it be said at once that this tautological transformation is extremely significant for our knowledge, and that the term 'tautological' implies nothing derogatory and is not a term of abuse.

With this view of thinking we stand opposed not only to rationalism and dualism but also to the view of some empiricists who believed they could derive logic (and mathematics) from experience by maintaining that the propositions of logic and mathematics expressed facts of experience – just like the propositions of physics – except that they dealt with especially frequent experiences, which is why these propositions acquired an especially high degree of certainty.

Indeed, the understanding of logic and mathematics has always been the main crux of empiricism; for any general proposition that has its origin in experience will always carry with it an element of uncertainty, whereas in the propositions of logic and mathematics we find no such uncertainty. We can imagine that the sun will not rise tomorrow, but we cannot imagine that $2 \times 2 = 5$, or that while $p \rightarrow q$ and p are true q is false; and the argument just mentioned – that this is a matter of especially frequent experiences – does not help us across this hurdle. *Only the elucidation of the place of logic and mathematics to be discussed below* (which is of very recent origin) *made a consistent empiricism possible.*

Observation and the tautological transformations of thought – these are the only means of knowledge we recognize. We do not recognize *a priori* knowledge, if only because we have no need of it anywhere: we do not know a single *a priori* synthetic judgement, nor do we know how it could come about; and as far as the so-called analytic judgements of logic (and mathematics) are concerned, these are directions for tautological transformations.

What is given is, for us, only what is individually perceived, what is immediately experienced by me, and only that has sense which is reducible in the final analysis to the given, which can be constituted out of the given. Thus the so-called 'permanent' objects of the external world known to us in daily life (tables, mountains, etc.) must first be constituted, and this holds also for the bodies of living things, of animals and men. Minds (one's own and others') must also be constituted – and experience teaches that they are attached to bodies; yet it would in principle be possible to constitute minds not attached to bodies; only so far it has not been shown to be necessary or even useful (contrary to the teachings of religion and spiritualism). The objects of physics (atoms, electrons, waves of all kinds) must also be reduced to the given by being constituted out of it. As soon as it appears impossible to constitute a term out of the given, the term must be regarded as senseless and along with it every proposition in which it occurs; and this is the fate of a large part of what is treated in philosophy, metaphysics, and theology.

The necessity of constitution shows already that in the acquisition of our knowledge we do not just receive the given but in addition process it. This provides us with an opportunity to sketch a view of the *place of logic* which is the subject of intense discussion within our circle. According to this view, logic is not something to be found in the given – or let us say: in the world. Logic is not, as used to be believed, a theory of the most general properties of objects, a theory of objects as such; rather, logic first arises when the given is processed, when the knowing subject confronts the given, tries to picture it to himself, and introduces a symbolism: logic is tied up with something's being *said* about the world. But this symbolism is not an isomorphic one-to-one projection – such a symbolism would be of very little interest, nor would the introduction of such a symbolism give rise to logic. Logic arises when and only when what is to be depicted and its pictures, the symbols, exhibit different structures. A very simple example of this is provided by negation itself: for every statement p there is a negation $\bar{p}$ in the symbolism, but only one of the two corresponding states of affairs occurs in the world; suppose it is the state of affair corresponding to p; we can then express it in two ways: by asserting p and by denying $\bar{p}$.

Take another example: besides the two symbols p and q, the symbolism contains the logical product $p \cdot q$ as a completely independent symbol; but *besides* the two states of affairs p and q, there does not exist in the world another independent state of affairs $p \cdot q$ coupled by a real 'and'. (The 'molecular' state of affairs $p \cdot q$ is already something imported into the given, the result of processing.) Now the symbol $p \cdot q$ makes it possible to assert p and q simultaneously; in saying $p \cdot q$ I say p along with it and likewise q. Logic arises from the circumstance that the symbolism we employ allows us to say the same thing in different ways, and that when we say something it allows us to say something else along with it. It is a set of directions on how something said can be said in another way, or how from something said we can extract another thing said along with it. This is what we will call the tautological character of logic.

And the occasion for introducing a symbolism whose structure deviates from what is to be depicted, a symbolism which allows us to say the same thing in different ways, is that we are not omniscient, that we can know the states of affairs that make up the world only in a very fragmentary way. This can again be seen very clearly in the example of negation: there would be no occasion for introducing negation if we knew every single state of affairs in the world. An omniscient subject needs no logic, and contrary to Plato we can say: God never does mathematics.

Hence logic is not a theory about the behaviour of the world –

on the contrary, a logical proposition states nothing at all about the world – but a set of directions for making certain transformations within the symbolism we employ. And once we take this view of logic, a much-discussed problem dissolves of its own accord – the problem of the seemingly mysterious parallelism between the course of our thought and that of the world, the seemingly pre-established harmony between thought and world, which would enable us to discover something about the world by thought. This is impossible in each and every case. Thought can only transform propositions tautologically and in so doing bring them into a form which we find easier to survey, or extract statements from them which had been asserted along with them and are more suitable for checking than the original propositions.

This is the role of thought in our system of knowledge and, in particular, it is the role of thought in physics; and a little reflection, a brief glance at the history of science shows that this role of thought – in spite of its tautological character – is not an incidental one. To take only one example, the basic equations of the electromagnetic field can hardly be checked immediately by observation; but by tautological transformation we recognize that in asserting them we assert statements about the situation which can be checked immediately, as is the case in Michelson's experiment, and this experiment can now serve as a check on the basic equations.

It was Wittgenstein who recognized the tautological character of logic and emphasized that there was nothing in the world corresponding to the so-called logical constants (like "and", "or", etc.), whereas Russell had taken the view that a statement was logical in character if it could be expressed solely by means of variables and logical constants; such a statement would be for example "there are twenty individuals in the world". But such a statement cannot be regarded as a logical one, since it asserts in fact something about the world and is in principle accessible to verification by experience.

Having said something about the place of logic in our system of knowledge, I must also say a few things about the place of *mathematics*. Since we have already adopted the view that experience and the logical transformations of logic are our only means of knowledge, and since nothing empirical enters into logic, the answer has been given in advance: mathematics must likewise have a tautological character, i.e., mathematics is part of logic. As is well known, this is Russell's position. But in adopting this position we do not mean to say that we regard *Principia Mathematica* as an infallible gospel; on the contrary, we believe that – in spite of the enormous merits of Russell and Whitehead – the task of developing mathematics as pure logic has not yet been completed – if only because the task of giving a satisfactory account of logic itself still awaits completion. Yet we are poles apart from the position of those critics who declare Russell's system to be finished and done with because there are still some difficulties in it, and who stand before his work just like a man confronted with the *Divine Comedy* when he knows no Italian and has access to only a brief summary of the poem.

Apart from a somewhat different view of logic itself, Russell's work seems to us in need of improvement as regards the theory of types it uses, which – it seems – ought to be replaced by a simpler one in which the highly controversial axiom of reducibility, which has been recognized, probably definitively, to be extralogical, becomes dispensable – and further, as regards the concept of absolute identity which Wittgenstein attacks – even though in this case too the remedy need not (it seems to me) be as drastic as Wittgenstein supposes.

Our adoption of Russell's position may cause some surprise in Germany where all ears are tuned to the controversy between intuitionism and formalism – between Brouwer and Hilbert. Yet it seems difficult for someone who has adopted the principles we have laid down to adopt Brouwer's *intuitionism*; its point of departure seems too much akin to Kant's pure intuitionism and Kant's *a priori*. This does not prevent one from rating very highly the internal mathematical significance of Bouwer's investigations: they establish what can or cannot be achieved in mathematics given certain means. The task of incorporating these investigations into the logicist system of mathematics is an important as well as urgent one, and required for the sake of completeness.

As regards Hilbert's *formalism*, what must be pointed out from our point of view is above all the unexplained role of metamathematical considerations. What is the origin of claims to metamathematical knowledge? It is certainly not experience! The difficulties we mentioned in speaking of attempts to base logic and mathematics on experience militate against this. Are they logical transformations? Certainly not! For they are supposed to serve as justifications for these transformations. What, then, are they? However, in adopting this sceptical position towards Hilbert's point of departure we do not mean to say anything against the significance of his investigations. On the contrary, I am convinced that many of Hilbert's concrete results will enter into the continuation and improvement of Russell's system.

What has been said so far about mathematics relates to arithmetic and analysis. The logicization of *geometry* was carried out in a very different way – with less controversy and at an earlier time. Its incorporation into logic was achieved by what is called axiomatization. Every theorem of geometry appears as the (tautological) implication $P \rightarrow Q$, where the antecedent P is the logical product of the axioms and the consequent Q the theorem in question. In this way the axioms no longer appear as self-evident though unprovable truths, but as assumptions from which deductions are made; and the basic concepts no longer appear as objects incapable of being dissected further by definition though capable of being grasped immediately by intuition, but merely as logical variables. Since every single axiom is a relation between the variables that represent the basic concepts, geometry appears as a special chapter of the theory of relations, as the investigation of certain special systems of relations.

But there is more to be said about geometry, a point that was all too often overlooked by mathematicians in their initial satisfaction with the logicization of geometry. Unlike logic and arithmetic whose propositions do not state anything about the world, geometry can also make a justified claim to being a factual science. It becomes one when its basic concepts, which on our view have so far been logical variables, are given an interpretation in reality, when they are constituted out of the given. Even if this constitution of the basic concepts of geometry does not belong to mathematics (which has nothing to do with reality), it is nevertheless an unavoidable task of scientific research. Following Whitehead, Russell took a giant step in this direction in Chapter IV of his book, *Our Knowledge of the External World*. It is, then, an empirical question whether the basic concepts constituted in this way satisfy the axioms; geometry in this sense is therefore an empirical science.

These last considerations bring us to physics. Some chapters of physics have already been axiomatized in the same sense as geometry and turned thereby into special chapters of the theory of relations. Yet they remain chapters of physics and hence of an empirical or factual science because the basic concepts that occur in them are constituted out of the given.

In doing this we may have the following goal in mind: to set up an axiomatic system by which the whole of physics is logicized and incorporated into the theory of relations. If we do this, it may well turn out that, as the axiomatic systems become more comprehensive, as they encompass more of the whole field of physics, their basic concepts become increasingly remote from reality and are connected with the given by increasingly longer, increasingly more complicated constitutive chains. All we can do is state this as a fact, as a peculiarity of the given; but there is no bridge that leads from here to the assertion that behind the sensible world there lies a second, 'real' world enjoying an independent being and differing in kind from the world of our senses, a world which we can never directly perceive, which manifests itself in the sensible world only in a distorted way, one difficult to unravel, and which we have to reconstruct from these sparse and confused clues. It seems to us that such a thoroughly metaphysical assertion is devoid of content, a mere pseudo-proposition; for the terms that occur in it cannot successfully be tied to the given by being constituted out of it.

And with this we have arrived again at the basic thesis of the scientific world view: there are only two means of acquiring knowledge: experience and logical thought; but the latter is nothing but tautological transformation, and therefore perfectly incapable of discovering a realm of being on its own, of leading outside the world of the given to a different kind of world of true being. Any kind of metaphysics is therefore impossible, and any metaphysical admixture is to be removed from science as a senseless combination of words.

The intrusion of metaphysical elements into science has been aided and abetted by two tendencies: a tendency to overestimate thought and to assume – while mistaking the tautological character of thought – that thought could lead all by itself to something new, and a tendency to overestimate *language*. Word language is a vey imperfect symbolism, its syntax accords badly with the syntax of logic; in particular, the subject–predicate structure prevailing in the languages we are familiar with and their partiality to nouns have caused much mischief in philosophy when it inferred the structure of the world from the structure of language, and they have led to the introduction of various pseudo-entities of a metaphysical character, such as substance, space, time, and number.

This defect of word language is one of the reasons why the adherents of the scientific world view also profess their adherence to what is called symbolic logic.

A further reason is this: the words of our language, besides referring to what they are supposed to symbolize according to their literal sense, carry along with them the most various kinds of accompanying images. Now these accompanying images encourage us to waver back and forth between the literal meaning and the 'metaphorical', 'figurative' meaning. Poetry rest almost entirely on this property of word language. But while poetry is quite justified as a means of expressing and generating feelings, poetry is also quite different in kind from the process of acquiring scientific knowledge whose essence is clarity and lack of ambiguity. And yet, by a careless and improper use of language many poetic elements have crept into science next to the many metaphysical elements – often hidden, shamefaced, and barely detectable, and certainly not always as obvious as in the following sentences, which I have taken from a treatise on 'Ontology' which appeared in Husserl's *Yearbook of Phenomenology*:

> The ecstasis of matter *posits* light. Where matter erupts from immanence into transcendence, it becomes lightlike in and through this process . . . . Where lightlikeness appears (on the other hand), we are dealing (as we know) with a real 'ecstasy'. What suffers the eruption is not the totality of matter within its outer confines, but the inner confinement of matter itself and as such. . . . The evident inner-directedness of the plain givenness of matter is transformed into the state of outer-directedness where matter *in itself* is lifted above itself.

In order to rid science of all 'poetic' elements, we need a symbolism whose symbols do not have the most various kinds of associations attached to them, as the words of our language do. If we try to imagine a poem written in the symbols of *Principia Mathematica*, we see at once how much symbolic logic is superior to ordinary language in this respect. And in this way the desire to free science from all metaphysical as well as all poetic admixtures turns the adherents of the scientific world view also into adherents of symbolic logic.

If it seems to a superficial observer that the scientific world view we have sketched stands opposed to the spirit of the times – which at present tends towards metaphysics, towards connections accessible only to mystical intuition and apprehensible only by feeling, and which aims at the whole, avoiding minutely detailed work – we are well aware that this is only a superficial judgement. The true expression of our time – a time of organizations with their stable structures, which owe their stability only to detailed work, a time of rationalization in industry, which operates within a well-defined system by taking in the smallest detail, a time of objectivity in architecture and the applied arts – the true expression of this time is the scientific world view with its loving, careful, detailed observation of the given, its cautious step-by-step logical constructions, and its plain language whose one and only task is to say clearly what is to be said.

DISCUSSION ABOUT THE FOUNDATIONS OF MATHEMATICS (*Sunday, 7 September 1930*)*

The following remarks are merely contributions to a discussion and thus necessarily mere sketches; I must therefore ask to be excused for not speaking with anything like the precision demanded by these questions.

If we want to decide in favour of one of the positions on the foundations of mathematics set out before us with supporting arguments, we must ask first of all: What are the requirements to be met by a theory of the foundations of mathematics? And in order to take a stand on this question, I must first preface a few words on philosophical matters.

It seems to me that the only possible position *vis-à-vis* the world is the *empiricist* one, which can very crudely be characterized as follows: any claim to knowledge which is not empty, which really says something about the world, can come only from observation, from experience; there is no way in which we can gain knowledge of reality by pure thought, and a single observation cannot furnish us with knowledge going beyond the individual case (the latter remark being directed against all doctrines of pure intuition and intuition of essences). I take this empiricist position, not because

* First published in *Erkenntnis* 2 (1931–2). This was Hahn's contribution to the discussion after R. Carnap, A Heyting, J. von Neumann, and F. Waismann had presented differing views on the foundations of mathematics. This was in the framework of the Second Conference on Theory of Knowledge in the Exact Sciences, which was held at Königsberg in 1930. Carnap's, Heyting's, & von Neumann's papers were also published in *Erkenntnis*. Some of Waismann's material may be found in his notes, *Wittgenstein and the Vienna Circle* (Oxford, 1979), pp. 102 ff.

I have selected it from among several possible positions, but because it appears to me to be the only possible one, because any real knowledge gained by pure thought, by pure intuition, by intuition of essences appears to me to be completely mystical.

Now if we try to work out this empiricist position, we seem to run up against a very simple fact: the fact namely that there is a logic and a mathematics which apparently furnish us with absolutely certain and universal knowledge of the world. This gives rise to the fundamental question: *How is the empiricist position compatible with the applicability of logic and mathematics to reality*? And in line with this question, the first requirement to be met by any theory of the foundations of mathematics is, to my mind, that it explain how the applicability of mathematics to reality is compatible with the empiricist position.

The representatives of intuitionism[1] and formalism[2] who have spoken to us have explained their positions so clearly that we can say quite definitely that neither intuitionism nor formalism fulfil this requirement. I regard the investigations of both Brouwer and Hilbert as highly significant within mathematics, but I do not regard them as theories of the foundations of mathematics. Mr. Heyting in his paper started out from a primitive intuition of the number series; to me there is something mystical about this primitive intuition, as there is about pure intuition or intuition of essences, and it is not therefore a suitable starting-point for the foundations of mathematics. And Mr. von Neumann has said as plainly as can be that formalism presupposes the whole of finite arithmetic, using it as its starting-point for justifying classical mathematics; but a position that presupposes finite arithmetic cannot be regarded as a theory of the foundations of mathematics.

Before explaining my own position, let me preface a brief argument: take any domain of objects, whatever they may be,

[1] A. Heyting [2] J. v. Neumann

with relations, whatever they may be, holding between these objects; let this domain be projected onto a pictorial domain in such a way that to the objects and relations of the original domain there correspond objects and relations of the pictorial domain; the objects and relations of the pictorial domain can then be conceived as *symbols* for the objects and relations of the original domain. If the projection chosen is not one–one but one–many, then one and the same state of affairs in the original domain will correspond to different combinations of symbols in the pictorial domain; there will therefore be transformations of combinations of symbols within this symbolism, and this gives rise to the task of giving rules for transforming one combination of symbols into another one which depicts the same state of affairs in the original domain. Now in my opinion the relation of language to reality is this: language correlates combinations of symbols with states of affairs in the world, and the way it correlates them is not one–one (which would be quite pointless) but one–many; and logic gives the rules about the way in which one combination of symbols in language can be transformed into another one which designates the same state of affairs; this is what is called the 'tautological' character of logic; a very simple example of it is provided by double negation: the proposition *p* and the proposition not-not-*p* designate the same state of affairs. Wherever there is a one–many projection, there is in this sense a 'logic' of this projection; what is ordinarily called 'logic' is the special case where we are dealing with the correlation of linguistic symbols with states of affairs in the world.

Logic does not therefore say anything about the world but has to do only with the way in which I talk about the world, and it should be evident that on this view the existence of logic is directly compatible with the empiricist position, whereas the conception of logic as a theory of the most general properties of objects is utterly incompatible with the empiricist position. Let us take for instance the logical principle $(x)\,\phi\,(x) \cdot \supset \cdot \phi\,(y)$, which says: what is true of all is true of each. This principle says nothing about the world; it is not a property of the world that what is true of all is also true of each; rather, the propositions "$\phi\,(x)$ is true of all individuals" and "$\phi\,(y)$ is true of each individual" are only different linguistic symbols for the same state of affairs; therefore the logical principle I cited only expresses a one–many relation of the symbolism we use as our language; it expresses the sense in which the symbol 'all' is used.

We now return to the foundations of mathematics. The logicist position explained by Mr. Carnap asserts that there is no difference between mathematics and logic. If this position can be made out, then the elucidation above of the place of logic within the system of our knowledge is also an elucidation of the place of mathematics, and the existence of mathematics is then also compatible with the empiricist position, like the existence of logic. And this is the reason why I opt for the logicist view among the three views on the foundations of mathematics laid out before us.

It can now be shown that the propositions of finite arithmetic, like $3 + 5 = 5 + 3$, have in fact the same tautological character as the propositions of logic: all one needs to do is go back to the definitions of the symbols 3, 5, +, and =. Finite arithmetic does not therefore present a difficulty to the logicist position. The situation is less clear as regards the transcendental methods of inference in mathematics, as in the theory of complete induction, in set theory, and in some chapters of analysis. There are some non-tautological principles which seem to play a part here: the axiom of choice for instance seems to have a real content, it seems really to say something about the world; at least this was Russell's position, and Ramsey's attempt to ascribe a tautological character to the axiom of choice was certainly a failure.

Russell's absolutist-realist position assumes that the world consists of individuals, properties of individuals, properties of such properties, etc.; and logical axioms are then supposed to be statements about the world. I have already said that this view is incompatible with a consistent empiricism, and I regard Wittgenstein's and the intuitionists' polemics against this view as fully justified; likewise I regard Ramsey's realist-metaphysical position, which has been attacked by Mr. Carnap, as impossible.

But while I do thus attack Russell's philosophical interpretation of his system, I nevertheless believe that the formal side of his system is largely in order as it is and highly suitable for the foundations of mathematics; we must only look for a different philosophical interpretation. Before I try to indicate such an interpretation, so as to make it easier to understand what follows I should like to refer you to a well-known fact. Think of any system of axioms for Euclidean geometry, e.g., Hilbert's. This system of axioms can be applied with extraordinary success in describing the world; and yet no one believes that it is possible to point out objects in the world that behave like the points, lines, and planes of Euclidean geometry; for we are here dealing only with idealizations, with assumptions that have been made for the purpose of giving a suitable description of the world.

I now assume, like Russell, that for describing the world (or better: a section of the world) we have at our disposal a system of predicative functions, of predicative functions of predicative functions, etc. – though, unlike Russell, I do not believe that the predicative functions are something absolutely given, something we can point out in the world. Now the description of the world will turn out differently according to the richness of this system of predicative functions; we therefore make certain *assumptions* about its richness; e.g., we will require that if $\phi(x)$ and $\psi(x)$ occur within the system, then so do $\phi(x) \vee \psi(x)$ and $\phi(x) \wedge \psi(x)$; we will also assume that, besides $\phi(x, y)$, $(y)\phi(x, y)$ also occurs within the system; we can also assume that the system is so rich that even a formula like $(\phi)\phi(x)$ will not take us outside the system; and the requirement that the axioms of infinity or the axiom of choice be valid is in this sense a requirement about the richness of the system of predicative functions by means of which I want to describe the world. Now the whole of mathematics arises out of the tautological transformation of the requirements we make about the richness of our system of predicative functions. Whether a certain proposition is or is not valid (e.g., the proposition about the cardinal number of the set of powers or the proposition about well-ordered sets) depends on the requirements we have made about the richness of the underlying system of predicative functions, or if you want to call them that, on the *axioms*; the question about the *absolute* validity of such propositions is completely senseless.

Someone may now perhaps want to raise the question: *Is there* such a system of predicative functions satisfying the above requirements? In the empirical sense (or in Russell's realist sense) there is certainly no such system: it is impossible to point out such a system in the world. Nor is there such a system in the constructive sense of the intuitionists. But nothing turns on this: just as Euclidean geometry is very useful for describing the world even though its points, lines, and planes cannot be exhibited, so the assumption of a system of predicative functions like the one discussed is very useful for describing the world even though such a system cannot be exhibited either empirically or constructively. Analysis, so construed, has only an hypothetical character: if I assume that for describing the world I have at my disposal a system of predicative functions satisfying certain requirements of richness, then the propositions of analysis are valid in such a description of the world. And in fact, the description of the world by means of analysis goes far beyond any possible empirical check. But we must, of course, presuppose that the requirements about the richness of the postulated system of propositional functions are *free from contradiction*, a point at which we make contact with Hilbert's ideas.

Now what is the meaning of an existential assertion in analysis as conceived here? It certainly does not assert any kind of constructibility in the intuitionist sense; but is it therefore as empty of meaning as the intuitionists think it is? Let us assume that some existential proposition has been proved by transcendental (and hence non-constructive) means, e.g. – to speak more concretely – the proposition, "There is a continuous function without a derivative", will anyone still attempt to prove the proposition, "Any continuous function has a derivative"? I believe not. And this shows that this bare existential proposition has a factual meaning; it does not mean that such a function is somehow empirically demonstrable in the world, nor that it is 'constructible'; rather, it has what I should like to call a 'technical' scientific meaning, that of a warning sign: Do not seek to prove the proposition, "Every continuous function has a derivative", for you will not succeed. That this is in fact the role of bare 'existential propositions' will – I think – be admitted by most of my colleagues who take an active part in the kind of research that is conducted, e.g., in the theory of real functions.

In conclusion I want to add a few words about Wittgenstein's criticism of Russell, which was the topic of Mr. Waismann's paper. I have already said that this criticism seems to me justified on some very important points. Yet I believe that the difference here is not nearly as big as it may seem from Waismann's paper. According to Russell, natural numbers are classes of classes; Wittgenstein's view seems to be a very different one; but if we bear in mind that Russell's symbols for classes are incomplete symbols which must first be eliminated if we want to recognize the real meaning of a proposition, and if this elimination is carried out according to the rules Russell gives, then we see that the two views are not so different after all. There is certainly the distinction that Wittgenstein emphasizes between a system and a totality, between an operation and a function, and it is correct to say that Russell does not make this distinction in his system.

However, operations and functions, systems and totalities have much in common and can therefore certainly be treated over a wide range of cases by employing the same symbolism. To criticize Russell effectively on this point, one would have to show that he goes too far in this common treatment, that he applies it also in cases where it can no longer be applied because of actual differences between the cases, and that he is thereby led into error.

EMPIRICISM, MATHEMATICS, AND LOGIC*

The fundamental thesis of empiricism is that experience is the only source capable of furnishing us with knowledge of the world, knowledge of facts, knowledge that has content: all such knowledge originates in what is immediately experienced. The place of mathematics has always presented a great difficulty for this position; for experience cannot provide us with *universal* knowledge, but mathematical knowledge seems to be universal; all knowledge originating in experience comes with a coefficient of *uncertainty* affixed to it, but in mathematics we notice no uncertainty. I know from experience that a stone left to itself will fall to the ground; but since I know this only from experience it can happen that my next trial will show it to be otherwise: that a stone left to itself will continue to float freely in the air; and we do not experience the least difficulty in visualizing this. If we knew only from experience that two times two is four, the same thing would have to be true of this proposition; but this is not the case: to imagine that two times two will tomorrow be five is completely impossible; we can be certain that this proposition was valid, is valid, and will be valid always and everywhere. And this is why any attempt to found the proposition "two times two is four" on experience is doomed to failure from the outset, and even the argument that this proposition rests on especially simple, especially frequent experiences and has therefore acquired such a high degree of certainty is of no help.

Thus it does in fact look at first sight as if pure empiricism was

* First published in *Forschungen und Fortschritte* **5** (1929).

bound to founder on the existence of mathematics, as if mathematical knowledge was knowledge of the world not originating in experience, as if it was a priori knowledge; and as is well known, this is in fact the opinion of some philosophers with purely rationalist tendencies and especially of philosophers within the Kantian tradition. And the difficulty this presents for empiricism is so striking and so radical that anyone who wants to advocate a consistent empiricism must face up to it. Let us make such an attempt here – by giving at least a few bare hints.

Let us remark first of all that what has been said about the propositions of mathematics is also true of the propositions of logic: the proposition "if all *a* are *b* and all *b* are *c*, then all *a* are *c*" is universally valid, like the proposition "two times two is four", and in this case too it is impossible to imagine that it might not be valid, say, tomorrow, and it too cannot therefore originate in experience. Let us then ask, to begin with: how can empiricism be made compatible with the existence of logic?

If logic were to be conceived – as it has actually been conceived – as a theory of the most general properties of objects, as a theory of objects as such, then empiricism would in fact be confronted with an insuperable difficulty. But in reality logic does not say anything at all about objects; logic is not something to be found in the world; rather, logic first comes into being when – using a symbolism – people *talk about the world*, and in particular, when they use a symbolism whose signs do *not* (as might at first be supposed) stand in an isomorphic one–one relation to what is signified (the introduction of a symbolism by means of an isomorphic one–one projection would be of very little interest). To take only one example: in *logic*, besides considering the statement *p*, we also consider its negation not-*p*; but in the *world* there always exists only one of the two states of affairs meant by *p* and not-*p*. Now our symbolism allows us to express the one state of affairs existing in the world in two ways: by affirming *p* and by denying not-*p*. If it were the case that both the state of affairs *p* and the state of affairs not-*p* existed in the world, and if the law of contradiction had as its content that the two existing states of affairs *p* and not-*p* were never to be encountered together in a certain way (to be further specified), then the law of contradiction would say something about the world and we should be faced with a dilemma fatal to empiricism: either the law of contradiction originates in experience, in which case it cannot be certain; or it is certain, in which case it cannot originate in experience. But since the two statements *p* and not-*p* do not correspond to two different states of affairs in the world but merely to *one*, which is only designated in different ways, the law of contradiction does not say anything about the world but deals, rather, with the way in which the symbolism used is supposed to *designate*.

And just like the law of contradiction, all the other propositions of logic do not say anything about the world either. Logic comes into being when the symbolism we use to talk about the world allows us to say the *same* thing in different ways, and the so-called propositions of logic are directions on how something we have said can also – within the symbolism used – be said in another way, or how a state of affairs we have designated in one way can also be designated in another way.

It would now be tempting to show how this conception of logic dissolves the seemingly remarkable problem of the parallelism between the course of our thought and that of the world – a parallelism that would put us in a position to discover something about the world by thought, as we seem to do in theoretical physics with such magnificent results. But we must resist the temptation and devote what little space is still available to returning to the problem with which we started: how is pure empiricism compatible with the existence of mathematics?

We have already indicated how pure empiricism is compatible with the existence of logic. The question "How is pure empiricism compatible with the existence of mathematics?" is therefore settled if it can be successfully shown that mathematics is part of logic, and hence, that the propositions of mathematics too do not say anything about the world, but are merely directions for saying what has been said in another way. As is well known, Kant launched the most decisive attack against the purely logical character of mathematics, and because of Kant's influence the efforts to dissolve mathematics into logic suffered a severe setback. But under B. Russell's leadership these efforts have recently been gathering enormous strength and now seem to be advancing on the road to victory. From the bare hints I have tried to give here in a few lines it will perhaps be gathered why empiricism in particular has a special interest in the purely logical foundations of mathematics. The investigations into the foundations of mathematics – seemingly the work of a few specialists – are in truth of enormous significance for the whole system of our knowledge: what is at issue in the dispute about the foundations of mathematics is the answer to the question: "Is a consistent empiricism possible?"

REFLECTIONS ON MAX PLANCK'S
*Positivismus und reale Aussenwelt**

The history of philosophy is pervaded by the doctrine – under a thousand different forms – that what we experience, see, hear, etc. . . . with our senses is mere appearance and that hidden behind it is a world of true being very different in kind and inaccessible to our senses.

This doctrine plays a large part in Indian philosophy. It arose independently in the very beginnings of Greek philosophy, and it has dominated European philosophy every since. The Eleatic doctrine of the One, Plato's theory of ideas, Leibniz's monadology, and the philosophy of Kant are all dominated by this view, and we encounter it again in the views of the most up-to-date physicists, like Max Planck, whose paper on "Positivism and the Real External World" will be discussed below. How was it possible for such a doctrine to arise? There were various reasons for it.

Take the phenomenon of dreaming. In dreams we experience all sorts of things similar to the things we experience while awake. But dreams are mere shadows: they have no substance, no reality corresponding to them; only what we experience in the waking state has a reality corresponding to it. But it is often extraordinarily difficult to decide whether something was really experienced or merely dreamt. Is our entire life perhaps nothing but a dream?

Closely related to dreaming are certain pathological states, e.g., delirium tremens: everything looks distorted, and some things are seen double; but this is only an illusion. Still, do we perhaps live constantly in such a state of delirium so that everything we see is

* Planck delivered this lecture in November 1930; it was published in Leipzig in 1931. Hahn's notes are a draft of the beginning of a reply to it. Schlick also wrote such a reply, "Positivismus und Realismus" (*Gesammelte Aufsätze*, The Hague, 1938 Chapter 5) [E. T. in *Synthese* 7 (1949)].

false? A very similar case is that of hallucinations. It is often extraordinarily difficult to decide whether something was hallucinated or really experienced. Many people swear to the reality of their hallucinations – e.g., people who have had visions of the saints – and it is impossible to talk them out of it. But the converse is also true: people in delirium tremens also imagine that they see mice, stag-beetles, and similar vermin.

Take the prodigies fakirs perform by suggestion. Are they real or not? Perhaps we are all under the hypnotic suggestion of a great fakir.

But disregarding dreams and pathological states, in the normal waking state we also constantly experience what are called 'deceptions of the senses'. A distant object looks smaller than close by, but in reality it is just as big. A stick dipped in water looks broken, but in reality it is straight. Or take a rainbow or a mirage. But if our senses sometimes deceive us, do they perhaps always deceive us? Perhaps everything is as unreal as a rainbow or a mirage.

Take a pair of distorting coloured glasses. Are we perhaps looking constantly through such glasses?

The question now arises: If we no longer trust our senses, what means have we of checking up on them; what means have we of apprehending a reality different from the world of our senses?

Many philosophers have been of the opinion that we do in fact have such a means at our disposal, namely thought. While our senses can deceive us and, evidently, often do deceive us, thought, in their opinion, provides us with an infallible means of exploring reality; it enables us to penetrate all appearance and to reach true being.

Thus Plato, to mention only one example, believed that he had to deny true being to the objects of the sensible world as a whole because their behaviour did not conform to our thought. It bothered him that there were many individual men but only one concept 'man' in our minds, that individual men came into being and passed away and underwent change, whereas the concept 'man' was forever unchangeable. All these deviations in the behaviour of individual men from the concept 'man' appeared to him as imperfections, and – believing that our thought is merely a likeness of true being and seeing that individual men behave differently from the concept 'man' – he arrived at the opinion that in the world of true being there must exist an entity corresponding exactly to our concept 'man', viz. the idea of man. And similarly for other concepts, e.g., numbers – and in this guise Plato's view has haunted philosophy even to this day.

The Eleatics were even more radical in this respect. They believed that any multiplicity, any limitation, any change, and hence also any motion was contrary to thought and therefore not to be encountered in true being, and since they are to be encountered everywhere in the sensible world, the sensible world must be mere appearance; being is one, unlimited, and unchangeable. Remember the flying arrow, to cite only one of the arguments.

This trust in thought, as opposed to distrust of the senses, may well have been due to the fact that, while the senses seemed deceptive, thought proved to be reliable in the affairs of daily life, as for example in the applications of mathematics where the results were always borne out by the facts – unless there had been an error of thought, i.e., a breach of the rules of thought. But in the course of time it became apparent that the far more extensive applications of thought in philosophy failed to deliver what they promised – there was no agreement to be had: what one philosopher extolled as an infallible result of exact thought, another declared to be sheer nonsense. And so the confidence in thought as a suitable means of advancing from the world of appearance into a world of true being came to be more and more discredited. To the representatives of our way of thinking it seems today to be an enormous misunderstanding of the nature and task of thought even to expect this sort of thing from thought. On our view, thought (meaning logical thought, which also embraces all mathematical thought and which alone enjoys demonstrative force) performs its operations only within language: we are dealing only with transformations which bring a proposition into a different form which says either the same or less; e.g., whether I say, "It is the case that *p*" or "It is the case that *p* or *q* and not *q*", I am saying exactly the same; and when I say, "It is the case that *p* and *q*", I am also saying, "It is the case that *p*". All thought is tautological. As soon as we are clear about this, we also see that thought cannot possibly do the job of bringing us news of a world of true being different from the world of the senses.

Now if it is true that the senses deceive us and that thought is not a means of penetrating to true being in spite of the deceptions of the senses, then we are apparently in a quandary.

But what about the assertion that the senses deceive us? Let us look once more at the arguments leading up to this assertion.

Take, e.g., the stick dipped in water: it looks broken to me, but in reality it is not broken, therefore my senses have deceived me. What do we mean when we say that it is not broken in reality? The facts of the matter are simply these: Under ordinary circumstances the optical appearance of a bent stick always occurs together with a certain tactile sensation: I can also feel the bend; but in the case of the stick dipped in water this is not so: while I have the visual impression of a bend, the tactile impression is that of a straight stick; so the tactile impression of a bend is absent. And for some reason or other we go more by the tactile sensation: when the tactile sensation of a bend is present we say that the stick is really bent, and otherwise, that it only seemed to be bent. In principle we should be equally justified in going more by the sense of sight and saying that when a stick is placed in water it is really bent, but that the sense of touch deceives us and does not reveal the bend. But the fact of the matter is simply that we are accustomed to having the visual and tactile impressions of a bend occur together, and are surprised when only the visual impression occurs (it is a peculiarity of ours not to be surprised when everything is as usual and to be surprised when it is not); but if we then say that the stick is not bent 'in reality', this is only a manner of expression which is due to the fact that we prefer the sense of touch, and all it means is that the tactile sensation of the bend is absent. It would be very much clearer if we had two different words, one for the visual sensation and one for the tactile sensation, e.g., 'stroped' and 'bent'; the state of affairs would then be expressed as follows: the 'stroped' sensation and the 'bent' sensation generally occur together; but when a stick is placed in water only the 'stroped' sensation occurs and not the 'bent' sensation; the ominous phrase 'in reality', which is liable to give rise to misunderstandings, has now disappeared (it was present earlier only because of the ambiguous use of 'bent', and hence, due to a defect in language), and we can no longer draw the conclusion: therefore our sense of sight has deceived us. All this is no more remarkable than that Italian women have generally black hair, but that once in a while we meet a blonde; from this we cannot conclude that our senses have deceived us – only, at best, that she has dyed her hair.

Take another example: when we sit in a train and it gradually begins to move, it often looks as if the train was at rest and the station-buildings had begun to move in the opposite direction. Here too it is customary to say that our senses have deceived us. What are the facts of the matter? What do we mean when we say that in reality the train is moving and the station-buildings are at rest? That can be stated in this case with fair precision: We mean that the train is changing its position relative to the earth and the buildings are not. Now the fact is that we directly perceive the change of place of an object with respect to us but not with respect to the earth (e.g., on the open sea where we see no land anywhere we have no idea whether a floating object is at rest relative to the earth or is moving with the current); but an object changes its place with respect to us not only when we move relative to the earth while the body is at rest but also the other way round. However, in the former case we are accustomed to gaining knowledge of our movement relative to the earth in other ways than just by the sense of sight, viz. by muscular exertion when we walk, and by bumps and noises when we ride. Now, when a train is gradually beginning to move and we do not notice any bumps or hear any noises, we conclude involuntarily that we are not in motion relative to the earth, and since we perceive the buildings changing their place relative to us, we conclude further that the buildings are in motion relative to the earth; hence the fact of the matter is not that our senses have deceived us, but that we have drawn – perhaps unconsciously – a false conclusion. The misleading phrase 'in reality' has again been entirely eliminated.

A similar case is that of the apparent motion of the stars. We say that the senses show the stars in motion; but this motion does not take place in reality, for in reality the earth is turning on its axis. In this case 'real motion' evidently means something different from the previous case, where it meant motion relative to the earth; for relative to the earth the stars *are* in motion. We therefore need to know first of all what is meant by 'real motion' in this case. And then it turns out that we cannot quite say what is meant by it in this case, at least not in a reasonably simple way; in fact, according to the theory of relativity the question is senseless in this form. The ominous phrase 'in reality' proves to be very dangerous. What it makes sense to say is that the stars show a change of place relative to the earth; but the question whether the earth is really moving but not the stars or the other way round does not make any sense, at least not without further data. If in this case too we draw the unconscious conclusion that, since we do not notice any muscular exertion, any shaking, or any noises, we are in reality at rest, this is due to the confusion of 'in reality', which is senseless in this case, with 'in reality' in the previous case where it meant 'with respect to the earth'. Once the phrase 'in reality' has been eliminated, there can be no talk of deception of the senses.

Let us now reflect on how it is in the case of the rainbow. We see it quite distinctly, at a very definite place. But if we now go towards its centre, it disappears, and there is nothing there except rain. Therefore – it is said – our sense of sight has deceived us; it showed us a coloured arch at a certain place, whereas in reality there was no coloured arch at that place. What is meant here by 'in reality'? When we see something at a certain place and go there, we are accustomed to having the visual impression remain while a tactile impression comes to be added to it. This is not so in the case of the rainbow. "In reality there is no arch" is then only supposed to mean that, contrary to what I am accustomed to, the visual impression of the rainbow vanishes as I approach it, and no tactile sensations occur at that place. It cannot be that my senses have deceived me. I have again concluded over-hastily that because I had this visual impression and because . . . , this visual impression must persist and be joined by a tactile impression. The fact of the matter is not that our senses have deceived us but that we have drawn a false conclusion. It is as if a boy accustomed to having his ears boxed when he sticks out his tongue does not have his ears boxed one day after sticking out his tongue and then says, in reality, therefore, I did not stick out my tongue.

Very similar things could be said about a mirage, a fata morgana, etc. In all these cases of what is called deception of the senses – all of which, incidentally, have long been incorporated into physical theories and have therefore nothing paradoxical about them for us – in all these cases it is like this: if we use the expression that we are dealing with 'deception of the senses', or that our senses conjure up something where there is nothing 'in reality', it is because from our sense impressions we have drawn over-hasty conclusions, which are not borne out by the facts. This becomes perfectly clear as soon as we try to give a concrete meaning to the phrase 'in reality', which in itself says nothing and is therefore misleading.

Once we make this clear to ourselves, it is not difficult to see how it is, e.g., in the case of an hallucination.

REVIEW OF ALFRED PRINGSHEIM,

Vorlesungen über Zahlen- und Funktionenlehre, Vol. I, Parts I and II, Leipzig and Berlin 1916*

Two parts of Volume I of Alfred Pringsheim's long-awaited *Lectures on Number Theory and the Theory of Functions* have now appeared: Part One is subtitled 'Real Numbers and Number Sequences', part Two 'Infinite Series with Real Members'; a third part will contain an introduction to complex numbers, the completion of the theory of series which this necessitates, and the theory of complex numbers and continued fractions. The second volume is to contain "an introduction to the theory of one-valued analytic functions of a complex variable and the simplest many-valued inverse functions on the basis of Weierstrass's methods and their further development, particularly with respect to the theory of integral transcendental functions and analytic progressions."

While some authors may give the title 'Lectures' to their works to indicate by way of excuse that their content is not quite well-rounded, not complete in itself, not quite finished, this is certainly not true of the present work. The lectures Pringsheim gave repeatedly at the University of Munich on this subject were already works of art completely ready for print and gone through with extreme care and full command over both form and content; and the present book, which came to be written "by summing up and sometimes further developing" these lectures, has attained a degree of perfection which – unfortunately but understandably – is very rare today in mathematical works. If in spite of this we do not, in reviewing this book, confine ourselves to a mere summary of it but take the occasion to also criticize it on some points, let it be

* First published in *Göttingische Gelehrte Anzeigen*, 1919.

stressed from the outset that – apart from occasionally pointing out some minor oversight or other, unavoidable in the first edition of a major work – we will be dealing almost exclusively with those very general questions concerning the *foundations* of the subject about which mathematicians are still far from unanimous, and about questions of a *methodological* nature which are to a certain degree questions of personal taste and which, as such, are perhaps always given different answers by different mathematicians.

To get down at once to a question concerning the foundations, let us start with a passage in the preface where it says, "It need hardly be mentioned that in spite of the elementary character of the presentation, I am striving for the greatest possible rigour in the proofs, for to my mind this ought to hold as a self-evident requirement for any mathematical presentation." One is struck by the fact that what is here required is merely the 'greatest possible' rigour, not *absolute rigour*; anyone who is acquainted with Pringsheim's works knows in advance that this is not meant to be a little back-door left open to let in inexactness at times when pedagogical considerations make it inconvenient or difficult to be rigorous. And yet that little phrase 'greatest possible' may not have been chosen by accident; perhaps it is meant to indicate that absolute rigour in the construction of number theory is a requirement which is very difficult to meet and which could not be satisfied at all within the given framework. For what is absolute rigour? When can we take a proof to be completely rigorous? Many will reply to this without a second thought: when it makes use only of purely logical means. But this raises the further question: what are purely logical means? A reference to the traditional textbooks of logic cannot be accepted as a satisfactory answer to this question, for if one thing is accepted as an incontrovertible fact by all mathematicians concerned with the foundations of their science, it is that the propositions of traditional logic are inadequate for the construction of mathematics, quite apart from the fact that the traditional presentations of logic are lacking in the kind of precision that would be required if they were to be used as the foundation for mathematical proofs. Therefore the mathematician who has taken to heart the logical construction of his science must characterize the traditional presentations of logic as materially inadequate and formally insufficiently precise. As is well known, this situation has led to attempts on the part of mathematicians to give logic the kind of shape needed by the mathematician for his purposes; let us refer here only to the system of *symbolic logic* worked out by Peano and his pupils. Now can the question raised above about purely logical means be answered by reference to Peano's *Formulary*? We are afraid that this too is not the case. We have reason to believe today that the incorporation of G. Cantor's set theory into the system of mathematics represents an ultimate and incontrovertible fact. But as is well known, certain contradictions appear in this discipline, due evidently to some defect in the logical foundations. But these contradictions appear not only if one makes use of traditional logic: Peano's logic too is incapable of avoiding these contradictions; but as long as it is incapable of doing this, it too cannot be taken to be an adequate logical foundation for mathematics.

Now it is true that a large-scale attempt has been made recently to shape symbolic logic in such a way as to avoid all contradictions and to present us with a complete table of all the fundamental concepts and propositions needed by mathematics for its construction.[1] I cannot pronounce judgement at this time as to whether this attempt has been successful; but one thing seems certain: this logical system has become so difficult and so comprehensive that practical considerations make it completely impossible to take it as the starting-point for a mathematical presentation of an elementary character and to answer the question raised above about the criteria of absolute rigour by reference to this system. While mathematicians since the days of Euclid have resigned themselves to the fact that there is no *royal road* to their science, access to it must not be so difficult that the only way to get to it is via the most tortuous and precipitous mountain trails, so that the majority of climbers will fall by the wayside, while the few who succeed in overcoming all difficulties will arrive exhausted to death – and not even at their goal but only at the beginning of mathematics proper.

It thus seems in fact that, as things stand at present, we cannot very well require absolute rigour and must be satisfied with the 'greatest possible' rigour, which, of course, leaves it to each person to decide what he will regard as the 'greatest possible'. This is of little significance once we have gone beyond the foundations, but as will be seen presently, it makes itself felt in rather an unpleasant way while we are still at the foundations. The author himself is evidently aware that what he regards as adequate rigour on these questions might not appear so to someone else; for in the preface itself he speaks of considerations which *to him* put the existence of natural numbers beyond doubt, "even at the risk that uncompromising logicians, axiomatizers, or set theorists might contradict me on this point".

If we must then perforce renounce the idea of going back *usque ad initium*, to the bitter beginning, in laying the logical foundations, where then are we supposed to start with the construction of mathematics? Most presentations, including also the present one, choose the concept of a *natural number* as their starting-point – and to me too this seems to be the natural thing to do. But this brings us again to a long-standing question which has kept philosophers interested in mathematics and mathematicians interested in philosophy in a state of suspense: can the theory of natural numbers (and hence also arithmetic and analysis) be constructed out of *purely logical* fundamental concepts and propositions, or does it require *specifically mathematical* fundamental concepts and propositions? Of course, this question has a precise sense only if we are given a table of those fundamental concepts and propositions that are purely logical. If we are forced to do without it, then our question can no longer be given a very precise answer. And so it happens that we cannot say with complete certainty what position the present work takes on this question.

Mr. Pringsheim has always been an effective and inspiring advocate of the position that the appeal to intuition is not an admissible means of conducting mathematical proofs; in his well-written works he himself has made important contributions to purging analysis of alogical, intuitive pseudo-proofs. It is true that he has been dealing with *geometrical* intuition; but what is good for geometry must be good for arithmetic: extralogical means of proof are inadmissible in arithmetic too. If arithmetic cannot dispense with extralogical elements, then it is in duty bound to set them out in the beginning as fundamental arithmetical concepts and propositions; as the source of the certainty of its fundamental propositions – if it feels in duty bound to inquire at all into this source – it may appeal to pure intuition or to some other source of knowledge; but any further construction must then be carried out with the exclusive help of the fundamental arithmetical concepts and propositions set out in the beginning and the purely logical ones. Various modern presentations have indeed taken this path, often using as their point of departure a system of fundamental concepts and propositions of arithmetic set up by Peano.[2]

Nowhere in the present work do we find an extralogical fundamental concept or proposition formulated and labelled as such. We can therefore hardly go wrong if we take the view that the author claims to be constructing arithmetic in a purely logical way; and we get additional confirmation for this from a passage in the preface where he emphasizes that he is letting the indefinitely continuable ordered sequence of natural numbers rise up, as it were, before the reader's eyes so as to put the existence of these numbers beyond doubt as far as the reader is concerned. This is followed by the passage mentioned above about being possibly contradicted by uncompromising logicians. To begin with, we must therefore take a closer look at the introduction of natural numbers.

The author starts with a simply ordered set satisfying the following requirements:[3] (1) The set itself and any subset arising from it by the omission of initial members have a first element.[4] (2) For any element except the first there is an immediately preceding element. (3) There is no last element. It is certain that as soon as the existence of such a set is established, the theory of natural numbers can be developed from it. Therefore the thing to do is to first of all prove the existence of such a set, which is best done by actually giving such a set. And this is then also the first aim the author pursues. He says that "the most primitive way [of producing such a set] is out of a single fundamental sign, say |, by successive repetition, as follows:

(*) |, ||, |||, ||||, |||||, . . . ,

but because this is extraordinarily imperspicuous it is completely useless". Instead of this procedure he therefore chooses another one which at the same time furnishes the decimal notation in Arabic numerals. But since we are concerned here only with questions of principle, and since our objection applies equally to both procedures but is more transparent when first applied to the former, let us return to the former. We cannot hide the feeling that there is a *petitio principii* here, concealed in the dots . . . that occur in (*). For if we wanted to state explicitly what these dots are supposed to indicate briefly, we should have to say: they indicate that the operation of adding element | is to be carried out according to type ω, i.e., these dots can be assigned a precise sense only if the type ordering the set of natural numbers is assumed to be known in advance.[5] Or expressed differently: by successive repetition of the fundamental sign | we can surely also produce a set of such fundamental signs ordered according to type ω (or according to the type of any transfinite ordinal number). But this is not what is meant here: the dots are supposed to indicate that the repetition of sign | can only be carried out a *finite* number of times, which makes it obvious that there is a *petitio principii* here.[6] Hence the attempt to construct a set with the properties postulated above seems to me to be a failure. But let it be emphasized at once that I do not regard this as a major disaster. Let us imagine the damage to have been repaired, by accepting the existence of a set of distinguishable things with the desired properties as an *unproved fundamental proposition*.

Now how do we get from the existence of such a set to the concept of a natural number? Here I must point out another difference of opinion between myself and the author, though this time it is not so much logical as more generally philosophical in nature. The author simply defines the elements of our set as natural numbers.[7] It is not completely clear to me what is meant by this. Mr. Pringsheim certainly does not take the view that the figure 1 I am now drawing with ink on paper is a natural number: it is obvious that he does not mean anything so concrete but something else, though it would not be easy to give a precise formulation to the underlying thought. On my view, the figure 1 can only be conceived as a conventional *sign* for the number one, which differs from signs like *a, b, x, y* only in this: that all men within our culture are agreed that (without an agreement to the contrary) such a sign is always supposed to mean the number one – just as they constantly designate by π the irrational number giving the relation of the circumference of a circle to its radius, without its occurring to anyone to say that the sign π *is* this irrational number. On my view, the situation is as follows: it can be proved that any two sets possessing all the properties required above are ordered by the same type, i.e., they can be projected onto each other in an isomorphic one-to-one manner. It can be proved further that between any two such sets there can be *only a single* isomorphic projection. This is why any element of such a set is correlated one-to-one with a certain element in any other such set. Any element of such a set stands therefore – from the purely ordinal point of view – in the same relation to its set as each of the elements correlated with it to its set. Now *this relation* is the natural (ordinal) number of the appropriate element.[8] The element itself can then be used as a *sign* for this ordinal number.

So much for the concept of a natural number as an ordinal number. But arithmetic cannot make do with this concept alone: it also needs the concept of a *cardinal number*. Without this concept we cannot even begin to formulate the general associative law. This state of affairs – which does not seem to have occurred to all authors who have written on the foundations of arithmetic – is formulated by the author with a clarity deserving of our gratitude; and since the concept of a natural ordinal number has already been introduced, the concept of a natural cardinal number must now be derived from it. As is well known, the basis for this derivation is the proposition: In the set N of natural numbers, no subset is equivalent to another subset or to set N itself. And it is essentially to this proposition that the reflections in Section 3 are addressed. But I cannot approve the claim that these reflections furnish a logical proof of the proposition in question. The line of thought is as follows: let 1, 2, 3, . . . , n be a subset S of N, and let $a, b, c, \ldots$ be the elements of this subset transposed into some other order. Imagine these elements in this order correlated one-to-one with the natural numbers 1, 2, 3, . . . Now let us first bring 1 back to its original place by transposition, then 2, and "continuing in this way we *finally* get to the point where each of the remaining elements 3, . . . , n has also been assigned a place marked by the corresponding number." Since during this procedure no place occupied by one of the elements $a, b, c, \ldots$ ever becomes vacant and likewise no initially vacant place ever becomes occupied, while in the final result the places 1, 2, . . . , appear to be all occupied, this must also have been the case initially; the numbers 1, 2, . . . , n were exactly correlated with subset S of N even when the order was changed to $a, b, c, \ldots$ This proof seems at first sight unobjectionable, and yet it can be shown by taking the following example that it cannot be compelling: Instead of subset S of N let us take set N itself, i.e., the set of natural numbers ordered according to type ω:

(*) 1, 2, 3, . . .

By transposing them as follows, we imagine them reordered according to type $\omega \cdot 2$:

(**) 1, 3, 5, . . . 2, 4, 6, . . .

We can now argue as above: let us first bring element 2 back to its original place by transposition, then element 3, etc. We thus obtain in sequence:

1, 2, 5, . . .	3, 4, 6, . . .
1, 2, 3, 7, . . .	5, 4, 6, . . .
1, 2, 3, 4, 9, . . .	5, 7, 6, . . .
. . .	. . .

In the course of these operations too, no place occupied in (**) ever becomes vacant; and here too we 'finally' (i.e., here, after an enumerable set of transpositions ordered according to type ω) get back to the original order (*) even though the elements in (**) are not correlated *in sequence* with the elements of (*). What is evidently in question is the meaning of the little word 'finally' which I emphasized above by putting it in quotation marks. The proof is compelling only if 'finally' is here taken to mean: 'by a *finite number* of transpositions'; but this also makes it clear that we are again dealing with a *petitio principii*, for the concept 'finite number' is yet to be introduced.

The point of this criticism is certainly not to declare that the way taken by the author is impassable, i.e., the way that is supposed to lead from the concept of a natural ordinal number to that of a natural cardinal number; this way is passable and even passable without difficulty. But one can also take the opposite way, which would agree more closely with the natural development of the number concept, as Pringsheim himself emphasizes. And what is more, in sharp contrast to Pringsheim I should like to regard the opposite way also as logically more convenient and more satisfactory. The concept of a cardinal number (or potency) as the invariant mark of a set under *any* one-to-one projection is for the set theorist a simpler concept than that of an ordinal number (or ordering type), which is an invariant mark of a simply ordered set only under a one-to-one projection which is also *isomorphic.*[9] If Mr. Pringsheim takes the contrary view and regards it "as a not very promising enterprise" to look for "a satisfactory elaboration of the theory of real numbers on the basis of the concept of a *cardinal number*" and even adds that "more recent, sometimes extremely artificial attempts of this kind" have confirmed his view to the fullest extent, I shall surely be permitted to indicate in a few words how such a theory can apparently be worked out in a perfectly natural way.[10]

Let us first define *unit sets* (in purely logical terms) by means of the following property: if a is an element of M and b is an element of M, then a is identical with b. Let us now define the cardinal number 1 as the potency of unit classes. Let us define further: where a is the potency of some set A, $a + 1$ is the potency of the conjunction of set A with a unit set not included in A; and finally let us define finite (or natural) cardinal numbers as those potencies that occur in any set which contains the number 1 and which, besides any potency a contained in it, also contains the potency

$a + 1$. This places complete induction at our disposal, and with its help we can easily prove that a set whose potency is the finite cardinal number n can be simply ordered only according to a single ordering type. These ordering types, correlated one-to-one with finite cardinal numbers in this way, are the finite (natural) ordinal numbers. I do not find that there is anything artificial about this line of thought. On the contrary, it seems to be the perfect purely logical clothing for the intuitive but imprecise ideas which everyone associates naïvely with the concepts of finite cardinal and ordinal numbers.

We have now examined in depth the lines of thought by which Pringsheim introduces the natural numbers, for we could not help noticing rather far-reaching differences between the author's views and our own on this point. But once we assume that we have reached the firm foundation of natural numbers on which further construction is based, we find that we can agree without reservation to nearly all of the subsequent developments. The few points on which we cannot completely follow the author concern individual questions of a methodological nature, of which we will only single out one for closer scrutiny, viz. the so-called *principle of permanence*. Let us look at what the author has to say about it in the preface (p. vii). After speaking of the introduction of fractions, zero, and negative numbers, he continues:

> To establish how these new numbers are to be *ordered* or, alternatively, *incorporated* into the ordered sequence of already existing numbers and how we are to *calculate* with them, we shall make use of the *principle of transference* which (following Haukel) is usually (but not very felicitously) called the principle of '*permanence*', and we shall make use of it in what I regard as a notably improved form which bestows on it the character of a certain logical necessity. For in every case we shall introduce *new number signs*, but only to such an extent that a subset of them represents signs for *already existing* numbers. The latter are therefore already governed by certain rules establishing their succession and defining the arithmetical operations for them, and these rules can without further ado be *transcribed* into the new

notation. If we are not to allow complete confusion in the manipulation of the total supply of newly created signs, we have hardly any choice but to extend the rules already governing part of it to the totality by definition, and to legitimize this step by proving that the stipulations we have made satisfy the requirements to be met by them without contradiction.

If we are to take a stand on these explications, we must preface a few words about the *principle of permanence*. Can this principle be somehow given a precise formulation? H. Schubert has made such an attempt in the *Encyclopaedia of the Mathematical Sciences*;[11] however, Peano has demonstrated convincingly that this attempt is a failure:[12] According to Schubert, the principle of permanence requires that the extensions of the number domain be carried out in such a way "that numbers in the extended sense are governed by the *same rules* as numbers in the non-extended sense". But this requirement *cannot possibly be met*. If it was really the case that numbers in the extended sense were still governed by the same rules, then they could not be distinguished from numbers in the non-extended sense, and we should not then be presented with an extension of the number domain. Hence it cannot be required that *all* propositions of the original number domain continue to hold in the extended domain; the most that can be required is that the most important propositions continue to hold. But what the most important propositions are is a matter of personal judgement. The principle of permanence thereby ceases to be logical in nature and becomes at best a piece of methodological advice containing within itself an element of arbitrariness. Now this element of *arbitrariness* has often been brought to the fore: it has been studiously maintained that we are making an arbitrary stipulation when in extending multiplication to negative numbers we let $(-1) \cdot (-1) = 1$, or when in extending the concept of an exponent we let $e^0 = 1$. Experience shows that students intellectually rebel against this: they have the feeling that there is something to the 'proofs' they were taught in school that $(-1) \cdot (-1) = 1$

or that $e^0 = 1$. And this gives us a hint that there is more to be done than just insisting on arbitrariness, that what is in question is not only an *element of arbitrariness* but also an *element of lawfulness* which must in turn be brought out into the open. Pringsheim's presentation too proceeds along this line, and I should like to declare that I agree without reservations with this tendency, but not quite with the way it is carried out. It looks to me as if Mr. Pringsheim himself had the feeling that his presentation does not yet amount to the last word: all he claims for it, according to the above quotation, is the character of "a certain logical necessity", and if this fatal word 'certain' becomes necessary at this point, it is, to my mind, because the elements of arbitrariness and lawfulness which cooperate in the usual extensions of the number domain are not as sharply separated as they might be. We will elucidate this by taking the example of the extension of the domain of natural numbers to that of positive rational numbers. Pringsheim's line of thought is as follows: A 'fraction' $\frac{b}{a}$ is to be regarded merely as a new sign for the natural number n whenever b is a multiple of a, or $b = na$. From this we obtain the proposition: two such 'improper' fractions, $\frac{b}{a}$ and $\frac{b'}{a'}$, are equal if and only if

(0) $\quad ba' = ab'$.

In the same way we obtain rules of addition and multiplication for such improper fractions – rules which can be proved as theorems. It then says (p. 41):

Now while improper fractions occurred merely as different signs for *natural numbers, proper* fractions[13] are perfectly *new signs* which we want to turn into new *number* signs by seeking to extend to them the relation of *succession* within the series of natural numbers ... and after that the fundamental operations of *addition* and *multiplication* – and we want to do all this in such a way that there arises no contradiction between the prior stipulations and arithmetical rules and the extended ones. If this goal can be achieved at all, then the possibility of success exists only if the rules for *improper* fractions which we obtained in the preceding paragraph as direct *consequences* of the rules governing natural numbers are now introduced as corresponding *definitions* of the relations of *proper* fractions among themselves and to *improper* fractions.

Now in this way we should indeed reach a "certain logical necessity"; we should now have no other choice (unless we wanted to renounce any attempt to extend the domain of numbers) but to define the equality of two fractions by means of (0), and all arbitrariness would be excluded.

But this is not, I am inclined to believe, how things are. While the conditions for the equality of two improper fractions can be expressed by (0), we have just as much right to express them by:

$$(00) \qquad (ba' - ab')^2 + (r_a(b) - r_{a'}(b'))^2 = 0,$$

where $r_a(b)$ (and analogously $r_{a'}(b')$) means the absolutely smallest residue of b to modulus a. If it were true that the domain of numbers *could* only be extended by assuming that (0) continued to hold, we should have just as much right to make the same assumption about (00). But this would lead to an entirely different definition of equality for improper fractions. Therefore it is evidently not the case that we may conclude without further ado that (0) gives us the only possible definition of equality for improper fractions. And if we prefer the definition of equality given us by (0) to the one given us by (00), which in and by itself is just as possible, it is not, it seems, because we are under a logical compulsion. Again, arbitrariness seems to reign supreme.[14]

May I now give a brief sketch of an alternative presentation which to my mind really separates what is arbitrariness from what is logical necessity and does full justice to each of these two elements.[15] Let me first specify the task: the system of natural numbers is to be extended by adding new 'numbers' in such a way that multiplication will always be one-to-one in the extended system. 'Multiplication' here is to be understood to mean a universally applicable associative and commutative connection which reduces to multiplication for natural numbers, as defined previously, when the two factors are equal to natural numbers. Our requirement that these properties of multiplication continue to hold (and not, e.g., the inequality $a \cdot b \geqq b$) is *arbitrary*; but from now on there is no longer any room for arbitrariness. Since, by stipulation, division is universally applicable within the extended system, the quotient of two natural numbers $b : a$ must be present within it; let us designate it by $\frac{b}{a}$; we then have:

$$(1) \qquad \frac{b}{a} \cdot a = b.$$

Now when are two signs, $\frac{b}{a}$ and $\frac{b'}{a'}$, to be defined as equal? If

$$(2) \qquad \frac{b}{a} = \frac{b'}{a'},$$

then also, by virtue of the concept of a one-to-one connection:

$$\frac{b}{a} \cdot (a \cdot a') = \frac{b'}{a'} \cdot (a \cdot a'),$$

and hence also, because of the associative and commutative character of multiplication:

$$\left(\frac{b}{a} \cdot a\right) \cdot a' = \left(\frac{b'}{a'} \cdot a'\right) \cdot a,$$

and therefore also, because of (1):

$$(3) \qquad b \cdot a' = a \cdot b'.$$

Conversely, if (3) is satisfied, then (2) follows from the requirement that division be one-to-one, because of

$$\frac{b}{a} \cdot (a \cdot a') = b \cdot a' \quad \text{and} \quad \frac{b'}{a'} \cdot (a \cdot a') = b' \cdot a.$$

We therefore have in fact no other choice but to define the equality of two fractions (2) by means of (3).

To find out how the multiplication of two fractions is to be carried out, we start with the relationship that follows from the properties required of multiplication and from (3), viz.:

$$\left(\frac{b}{a} \cdot \frac{b'}{a'}\right) \cdot (a \cdot a') = \left(\frac{b}{a} \cdot a\right) \cdot \left(\frac{b'}{a'} \cdot a'\right) = b \cdot b'.$$

This shows that if division is to be carried out one-to-one as required, we have no other choice but to define:

$$\frac{b}{a} \cdot \frac{b'}{a'} = \frac{b \cdot b'}{a \cdot a'}.$$

And now we can go on to *prove* that there is one and only one distributive connection for multiplication, which reduces to the addition of natural numbers when the two connected fractions are equal to natural numbers. Thus if we demand that the addition of fractions should also be distributive (a demand which is again arbitrary), then we must of necessity define it by:

$$\frac{b}{a} + \frac{b'}{a'} = \frac{a'b + ab'}{aa'}.$$

I hope that by this brief development of my views I have made it clear enough what it is about Pringsheim's presentation of successive extensions of the number domain that appears to me not quite satisfactory and in what direction his presentation seems to me capable of improvement.

I should still like to touch on two questions of principle before I turn to a brief summary of the contents of the book under review. Mr. Pringsheim is today the most eminent representative of the arithmetical school, a school which goes so far in rejecting any geometrical element in analysis that it even renounces the help of the highly suggestive geometrical *terminology* which would make its propositions and proofs easier to understand. Nobody will therefore be surprised if he does not find even a trace of geometry in Pringsheim's *Lectures*. Since these lectures may well have to count for a long time to come as the classical representation of the arithmetical school in its purest form, we must also, in reviewing this work, take a stand on the question of the complete banishment of everything geometrical from analysis.

Everyone knows how the great process of the arithmetization of analysis was started in the last century and carried almost to completion: Mathematicians had for a long time made use of supposedly geometrical evidence as a means of proof in much too naïve and much too uncritical a way, till the unclarities and mistakes that arose as a result forced a turnabout. Geometrical intuition was now declared to be inadmissible as a means of proof, and what was demanded was complete 'arithmetization'. The slogan of arithmetization does not seem to me to be a felicitous one, or at least it does not seem to me to express what is required today. As there is *geometrical* evidence of an alogical nature (this is what is called 'intuitive' evidence), so there is, no doubt, *arithmetical* evidence of an alogical nature (perhaps I may call this evidence too 'intuitive'): thus we come to know, e.g., the commutative property of addition for natural numbers with all the evidence we need even without logical analysis. Now if, without further justification, one excludes geometrical intuition as a means of proof but admits arithmetical intuition, this seems to me dogmatic and arbitrary. The ideal, therefore, is not arithmetization but *logicization*, and in this form the requirement holds equally for arithmetic, analysis, and geometry. Logicization is achieved when the whole discipline is developed from a series of fundamental concepts and propositions by purely logical means; it is even more fully achieved when the discipline requires nothing but purely logical fundamental concepts and propositions for its development.

Now by what right could analysis from a purely logical standpoint reject a geometry logicized in this sense? Surely only if such a geometry needed some fundamental concepts or propositions for its construction which were not needed by analysis. Now is this the case? Obviously not. This is evident where geometry rests on the kind of foundations E. Study lays down in his book, *The Realist World View and the Theory of Space,* [16] where for instance a point in a plane is *defined* as an ordered pair of real numbers;

but it is no less true of the so-called axiomatic systematizations of geometry, where geometrical concepts are implicitly defined by the system of 'axioms',[17] and where the proof of the existence and uniqueness of these concepts is given subsequently by an appeal to analysis.[18] Taken *purely logically*, the banishment of all geometry from analysis does not therefore seem to me justified.

Indeed we seem to be dealing here with a *psychological* or *pedagogical* question: it is feared that the use of geometrical concepts might bring with it an irresistible temptation to use (perhaps unconsciously) intuitive and alogical means of proof. That there is such a danger must definitely be admitted. But whether this advantage is outweighed by the disadvantages, which to my mind are quite significant, will always remain a matter of opinion. I regard these disadvantages as so significant because I am convinced that all progress in the more subtle parts of analysis, as for instance in the theory of real functions, is first achieved subjectively, in an intuitive way, and hence that the use of intuition is an indispensable means of investigation. Of course, I do not mean the kind of crude and uncritical use by which analysis was led astray at one time, but a use of intuition purified and refined by the knowledge gained over a century. It seems to me an important task of mathematical instruction, in the classroom as well as in textbooks, to go on strengthening and refining this means of investigation – without damage to the logical rigour which is an inexorable requirement of all proof under all circumstances – instead of letting it atrophy by casting it aside.

So much for the first question of principle on which I was going to take a stand. The second question concerns voluntary confinement to '*elementary methods*'. As the author writes in the preface:

> And as the fundamental idea which was constantly before my mind and guided me in the composition both of the arithmetical part and of the part dealing with the theory of functions, I should like to single out the idea of carrying

out the intention to *take advantage* as much as possible of *elementary methods* and to *develop* them far enough so as to be able to familiarize the reader with modern means of sharpening and deepening the concepts and questions, and to lead him, as closely as the simplicity of the methods employed will allow, to the limits of our present knowledge.

I have often asked myself in vain for a definition of 'elementary' methods; and I was hoping to find it in this work; but my hopes were disappointed, at least in the instalments we have so far. In asking for such a definition I am not, of course, asking for a mere enumeration of those methods that are to count as elementary, but rather for a statement of the principles by virtue of which one method is to be regarded as elementary and may therefore be used, whereas another is to be excluded from use because it is not elementary. What is the characteristic difference, e.g., between the methods of approximation implicit in the concept of a power series and in that of an integral, which in effect allows the former but not the latter to serve as the foundation of an 'elementary' theory of functions? But as long as such a characteristic difference between elementary and non-elementary methods has not been established, many will hardly be able to suppress the feeling that there is a certain arbitrariness and, one might almost say, a kind of self-mutilation in such voluntary confinement to elementary methods, and that one ought not to sacrifice the simplicity even of a single proof for their sake. On the other hand it cannot, of course, be denied that such confinement in the choice of methods achieves a certain aesthetic effect, a certain harmonious uniformity, which is present in a high degree especially in the present work. And we should therefore only like to express the wish that the questions of principle attached to the term 'elementary methods' will soon be completely clarified. No one would be more competetent to make a contribution to such clarification than the most eminent and most effective advocate of these elementary methods, namely Mr. Pringsheim himself.

[The remainder of the review – a summary and comments on some points of purely mathematical interest – is here omitted].

NOTES

[1] A. N. Whitehead and B. Russell, *Principia Mathematica*, Cambridge.

[2] G. Peano, *Arithmetices principia nova methodo exposita*, Turin 1889. This system is also found in appendix II of A. Genocchi and Peano, *The Differential Calculus and the Principles of the Integral Calculus* [*Calcolo differenziale* &c (Rome, 1884) G. T. *Differentialrechnung* &c. (Berlin, 1889)].

[3] In the book it says: "an endless sequence". I am inclined to believe that what is understood by the term 'sequence' of things is, quite generally, the occupancy of the ordered set of natural numbers by these things, so that I prefer to avoid the word 'sequence' at this point where the concept of a natural number remains to be developed.

[4] That is, in the usual terminology of set theory: the set is *well-ordered*.

[5] We have to raise the very same objection against A. Loewy's presentation of the theory of natural numbers in his *Lehrbuch der Algebra*, which appeared recently. Loewy (on p. 2) takes Peano's axioms as his basis, but replaces Peano's fifth axiom, "if class s contains the number 1 and, besides any number x that occurs in it, also the (immediately following) number x^+, then it contains class N of all natural numbers", by the axiom: "Any element of N is contained in the system $1, 1^+, 1^{++}, 1^{+++}, \ldots$" To my mind this alteration causes Peano's system to lose all of its sense. In the latest edition of O. Stolz and J. A. Gmeiner, *Theoretische Arithmetik*, Peano's axioms have also had to suffer some alterations, which deal a fatal blow to this well-thought-out and clearly-formulated system: in Gmeiner's reformulation of these axioms (on p. 15) there is talk of "the numbers following a number a in the number sequence", a concept which is elucidated by means of a^+, $(a^+)^+$, etc. Here we have again the ominous 'etc.', which has no sense at this point where the concept of a natural number or – if you like – of type ω is not yet available and which therefore contains a *petitio principii*. However, when Peano's axioms are reproduced correctly, it is possible to *deduce* an order of natural numbers from them and thereby to attach a precise sense to the concept 'the numbers following number a'.

[6] Against this it must not be objected that it is impossible to repeat a process infinitely many times. It is also impossible to repeat it $10^{10^{10}}$ many times. But if someone wanted to assert that a process cannot be repeated infinitely many times in a different sense and, as it were, in a higher degree, differing from the sense in which it cannot be repeated $10^{10^{10}}$ times, he would be in duty bound to back this up, i.e., to explicate the precise sense of the terms

'can' and 'cannot be repeated', which would lead to extremely difficult questions of a psychological, epistemological, and even metaphysical nature, which he will definitely want to avoid in the foundations of mathematics.

7 For the sake of accuracy I should not like to leave it unmentioned that Pringsheim does not speak of a set of any elements whatever but demands of these elements that they be 'signs'. But now, any thing can surely be used as a sign for any other thing, so that it seems to make no difference whether we speak of sets of arbitrary elements or of sets of signs. It certainly makes no difference for the discussion of the text.

8 If we laid down only the first of the three requirements for our sets, we should also obtain Cantor's *transfinite ordinal numbers*. There can be no such purely ordinal definition of *rational* (or of *real*) numbers, because two sets whose ordering type is that of the naturally ordered rational numbers can be projected onto each other, not just in *one* way, but in infinitely many ways.

9 Symbolic logic does not distinguish between the mark characteristic of all things of a class and the class itself. And since it employs the words 'set' and 'class' synonymously, it used to define the potency of class A simply as the class of all classes equivalent to A. This definition has been abandoned because the concept of the class of all classes equivalent to A proved to be self-contradictory. I believe that this definition can be retained if we make a minor modification in it. Let us start by supposing that we are given a domain D of things. Let us designate collections of these things as sets and lay it down, as a fundamental logical law, that a set is not a thing in D. Let us now extend domain D to D' by adding sets of improper things. We can now form sets of things in D'; to distinguish them from the previous sets, let us call them 'second-level sets' and the previous ones 'first-level sets'. The definition of a potency will then read: The potency of a first-level set A is the second-level set of all first-level sets equivalent to A. As far as I can see, this reading no longer gives rise to contradictions. Cf. M. Pasch, *Grundlagen der Analysis*, p. 94.

10 Cf. B. Russell, *The Principles of Mathematics*, p. 128.

11 *Encyklopädie der mathematischen Wissenschaften*, Vol. I, Part 1, p. 11.

12 *Revue de mathématique* 8 (1903).

13 These are fractions in which the numerator is not a multiple of the denominator.

14 It should be obvious that the concept 'arbitrariness' is to be understood throughout this discussion in a purely logical sense, and hence merely as the contrary of the concept 'logical necessity'. It may very well happen that there are several cases each of which is logically possible, so that a choice between them is *logically* arbitrary, but only a single case which is *practically* possible for reasons of realizability or applicability.

15 The general theory of the extension of a system of magnitudes, as pre-

sented by O. Stolz and J. A. Gmeiner in *Theoretische Arithmetik*, Section 3, Paragraph 7 (p. 67), could be reformulated in exactly the same way.

16 *Die realistische Weltansicht und die Lehre vom Raume*, Chapter V, p. 81ff.

17 These 'axioms' are therefore far from being 'fundamental propositions' in the sense mentioned above; nor is the requirement that a particular connection be associative and commutative a 'fundamental proposition' when it is part of a study of the most general associative and commutative connections in a system of magnitudes (as conducted, e.g., by Stolz and Gmeiner in their *Theoretische Arithmetik*, 2nd ed., p. 50ff.).

18 For purposes of analysis, the first-mentioned view will have to be given preference over the axiomatic one, because the latter is methodologically, though not logically, alien to analysis, which is not true of the former. Anyone who refuses to accept the name 'point' for an ordered pair of real numbers would also, for the sake of consistency, have to refuse to accept 'real number' as a proper name for a convergent series of rational numbers.

THE CRISIS IN INTUITION*

Of all the leading philosophers Immanuel Kant was undoubtedly the one who assigned the greatest importance to the part played by intuition in what we call knowledge. He observed that two opposite factors are basic to our knowledge: a passive factor of simple receptivity and an active factor of spontaneity. In his *Critique of Pure Reason*, at the beginning of the section entitled 'Transcendental Theory of Elements; Part Two: Transcendental Logic,' we find the following:

"Our knowledge comes from two basic sources in the mind, of which the first is the faculty of receiving sensations (receptivity to impressions), the second the ability to recognize an object by these perceptions (spontaneity in forming concepts). Through the first an object is *given* to us, through the second this object is *thought* in relation to these perceptions, as a simple determination of the mind. Thus, intuition and concepts constitute the elements of all our knowledge . . . ".

That is to say, we conduct ourselves passively when through intuition we receive impressions, and actively when we deal with them in our thought. Further, according to Kant, we must distinguish between two ingredients of intuition. One of these, the empirical, *a posteriori* part, arises from experience and forms the *content* of intuition, such as colours, sounds, smells, and sensations of touch (hardness, softness, roughness, etc.). The other is a pure, *a priori* part, independent of all experience; it constitutes the *form* of intuition. We possess two such pure intuitional forms: *space*, the intuitional form of our external sense by means of which we

* First published in *Krise und Neuaufbau in den exakten Wissenschaften*, Fünf Wiener Vorträge, Leipzig and Vienna, 1933.

"picture things as outside ourselves"; and *time*, the intuitional form of our inner sense "by means of which the mind observes itself or its inner state."

In Kant's system, as I have said, this pure intuition plays an extremely important role. He believed that mathematics is founded on pure intuition, not on thought. Geometry, as it has been taught since ancient times, deals with the properties of the space that is fully and exactly presented to us by pure intuition; arithmetic (the study of the real numbers) rests on our pure and fully exact intuition of time. The intuitional forms of space and time constitute the *a priori* frame into which we fit all physical happenings that experience presents to us. Every physical event has its precise and exactly determined place in space and time.

However plausible these ideas may at first seem, and however well they corresponded to the state of science in Kant's day, their foundations have been shaken by the course that science has taken since then.

The physical side of the question has already been treated in the first two lectures,[1] so I can here confine myself to mentioning it briefly. Kant's ideas about the place of space and time in physics correspond with Newtonian physics, which was supreme in Kant's day and which remained so down to very recent times. This conception received its first serious jolt from Einstein's theory of relativity. According to Kant, space and time have nothing to do with each other, for they stem from quite difference sources. Space is the intuitional form of our outer sense, time of our inner sense. We have an absolutely stationary space and an absolute time that flows independent of it. The theory of relativity holds, on the contrary, that there is no absolute space and no absolute time; it is only a combination of space and time – the 'universe' – that has absolute physical meaning.

A much worse blow was struck at Kant's conception of space and time as *a priori* intuitional forms by the most recent developments of physics. We have already noted that, according to Kant's conception, every physical event has its precisely fixed location in space and time. But there has always been a certain difficulty about this. We know physical events only through experience: but all experience is inexact and every observation involves observational errors. Thus the earlier conception embodies the inconsistency that, while every physical event has its exact place in space and time, we can never precisely determine those places. Let us take, for example, a circular piece of chalk: once a unit of length is chosen, the distance between any two points on this piece of chalk is measured by an exact real number. Imagine that the distance has been determined between every pair of points on this piece of chalk and call the greatest of these distances its 'diameter'. Assuming that the chalk occupies an exact fixed portion of the space given to us by precise intuition, it would be reasonable to ask, "Is the diameter of the chalk disc expressed by a rational or an irrational number?" But the question could never be answered, for the difference between rational and irrational is much too fine ever to be determined by observation. Thus what might be called the classical conception raises questions that are fundamentally unanswerable: which is to say that the conception is metaphysical.

For a long time this difficulty was not taken very seriously. It was answered somewhat as follows: "Even if every single observation is inexact and subject to observational errors, yet our methods of observation are becoming more and more accurate. Let us now imagine a specific physical quantity measured over and over again with more and more precise observational methods. The results thus obtained, though each one is inexact, will nevertheless approach without limit a definite limiting value, and this limit is the exact value of the physical quantity in question." This argument is scarcely satisfactory from a philosophical standpoint, and the recent advances of physics seem to prove that it is also untenable on purely physical grounds. For it now seems that on purely physical grounds the location of an event in space and time cannot be determined with unlimited precision.[2] If, then, measurements cannot be pushed beyond certain limits of exactness we are left with this result: The doctrine of the exact location of physical events in space and time is metaphysical, and therefore meaningless. This most recent and revolutionary development of physics will necessarily come as a shock to most persons – including most physicists – grounded as they are in dogmatic and metaphysical theories; but for the thinker trained in empirical philosophy it contains nothing paradoxical. He will recognize it at once as something familiar and will welcome it as a major step forward along the road toward the 'physicalization' of physics, toward cleansing physics of metaphysical elements.

After this very brief reference to the physical side of the question we turn to the field of mathematics where the opposition to Kant's doctrine of pure intuition manifested itself considerably earlier than in physics. From here on I shall deal exclusively with the subject "mathematics and intuition"; moreover, even within this sphere I shall pass over a whole group of questions, as important as they are difficult, which Menger will deal with in the final lecture of this series.[3] I shall not discuss the vehement and successful opposition to Kant's thesis that arithmetic, the study of numbers, also rests on pure intuition – an opposition inextricably bound up with the name of Bertrand Russell, and which has set out to prove that, in complete contradiction to Kant's thesis, arithmetic belongs exclusively to the domains of the intellect and of logic.[4] Thus I have narrowed my subject to "geometry and intuition," and I shall attempt to show how it came about that, even in the branch of mathematics which would seem to be its original domain, intuition gradually fell into disrepute and at last was completely banished.

One of the outstanding events in this development was the

discovery that, in apparent contradiction to what had previously been accepted as intuitively certain, there are curves that possess no tangent at any point, or (and we shall see that this amounts to the same thing) that it is possible to imagine a point moving in such a manner that at no instant does it have a definite velocity. Mathematicians were tremendously impressed when their great Berlin colleague C. Weierstrass made this discovery known in the year 1861. But manuscripts preserved in the Vienna National Library show that the fact was recognized considerably earlier by the Austrian philosopher, theologian and mathematician, Bernard Bolzano. Since some of the questions involved here directly affect the foundations of the differential calculus as developed by Newton and Leibniz, I shall first say a few words about the basic concepts of that discipline.[5]

Newton started with the concept of velocity. Imagine a point moving along a straight line, as shown in Figure 1. At time t the moving point

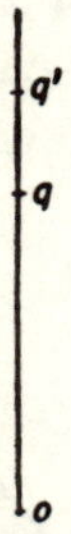

Fig. 1.

will be, say, at q. What is to be understood by the expression "the velocity of the moving point at the instant t"? If we determine the position of the point at a second instant t' (at this second instant think of the point as being at q'), then we can ascertain the distance qq' that it has traversed in the time that has elapsed between the instants t and t'. We now divide the distance qq' that the point has traversed, by the time that has elapsed between the instants t and t', and get the so-called 'mean velocity' of the moving point between t and t'. This 'mean' velocity is in no sense the velocity at time t itself. The mean velocity may, for instance, turn out to be very great, even though the velocity at instant t is quite small, if the point moved very rapidly during the greater part of the time interval in question. But if the second instant t' is chosen sufficiently close to the first instant t, then the mean velocity between t and t' will provide a good approximation to the velocity at time t itself, and this approximation will be closer, the closer t' is to t. Newton's reasoning about this matter ran somewhat as follows: Think of the instant t' chosen closer and closer to t; then the average velocity between t and t' will approach closer and closer to a cêrtain definite value; it will – to use the language of mathematics – tend toward a definite limit, which limit is called the "velocity of the moving point at the instant t." In other words, the velocity at t is the limiting value approached by the average velocity between t and t', as t' approaches t without limit.

Leibniz started from the so-called tangent problem. Consider the curve shown in Figure 2; what is its slope (relative to the horizontal) at point p?

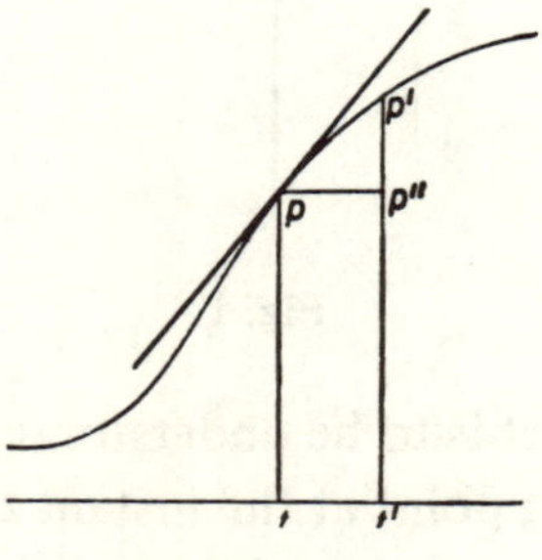

Fig. 2.

Choose a second point, p', on the curve and construct the 'average slope' of the curve between p and p'. This is obtained by dividing the height (represented in Figure 2 by the line $p''p'$) gained in ascending the section of the curve from p to p' by the horizontal projection of the distance passed over (represented in Figure 2 by the line pp'', which indicates how far one moves in the horizontal direction by following along the section of the curve from p to p'). The average slope of the curve between p and p' is not, of course, identical with its slope at the point p itself (in Figure 2 the slope at p is obviously greater than the average slope between p and p'). However, it will give a good approximation to the slope at p, if only p' is chosen sufficiently close to p; and the approximation will be more accurate, the closer p' is to p. Now again as in the Newtonian example: If p' is permitted to approach p without limit, the average slope of the curve between p and p' will tend toward a definite limit, which limit is called the "slope of the curve at the point p." That is to say, the slope at p is the limiting value approached by the average slope between p and p' as p' approaches p without limit. One designates as the "tangent of the curve at the point p" the straight line passing through p which (throughout its entire length) has the same slope as the curve at p.

There is thus a striking resemblance between the procedure for obtaining the slope of a curve and the procedure for determining the velocity of a moving point. In fact, the problem of determining the velocity of a moving point at a given instant becomes identical with the problem of determining the slope of a curve at a given point if we employ a simple device, familiar[6] from its use in graphic railway timetables. Along a horizontal straight line (a "time axis") mark off time intervals so that every point on the line represents a definite point of time; and on the straight line in Figure 1, along the path of the moving point in question select an arbitrary point o. If at instant t the moving point is at q, erect at right angles to the time axis at t the line segment oq. (See Figure 2.) It can be seen that the point p thus obtained will represent as in Figure 2 the position of the moving point at instant t. If this process were carried out for every single instant of time, one would obtain a smooth curve portraying the path of the moving point, namely its 'time–distance curve'. From this curve one can derive all the particulars of the motion of the point, just as one can work out a train schedule from the graphic type of timetable referred to above. Now it is evident that the average slope of the time-distance curve between p and p' is identical with the average velocity of the moving point between t and t', and thus the slope of the time-distance curve at p is identical with the velocity of the moving point at the instant t. This is the simple connection between the velocity problem and the tangent problem; the two are the same in principle. The fundamental problem of the differential calculus is this: Let the path of a moving point be known; from these data its velocity at any instant is to be calculated; or, let a curve be given – for each of its points the slope is to be calculated (at every point the tangent is to be found). We shall now examine the tangent problem, bearing in mind that everything we say about this problem can, on the basis of the foregoing, be carried over directly to the velocity problem.

We noted that if the point p' on the curve in question approaches the point p without limit, the average slope between p and p' will approach more and more closely a definite limiting value, which will represent the slope of the curve at the point p itself. It may now be asked whether this is true for every curve. The principle holds for the standard curves that have been studied since early times: circles, ellipses, hyperbolas, parabolas, cycloids, etc. But a relatively simple example will show that it is not true of every curve. Take the curve shown in Figure 3; it is a wave curve, and in the neighborhood of the point p it has infinitely many waves. The wave length as well as the amplitude of the separate waves decrease without limit as they approach p. Using the method described above we shall attempt to ascertain the slope of the curve at the point p. We take a second point p' on the curve and

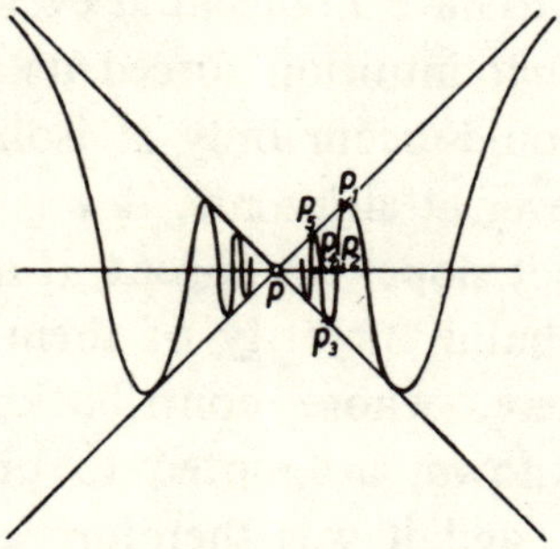

Fig. 3.

obtain the average slope between p and p'. If we take p_1 as the point p' (Figure 3) the average slope between p and p_1 turns out to be equal to 1. If p' is permitted to approach p along the curve, it can be seen that the average slope at once decreases; when p' reaches p_2 the average slope (between p and p_2) becomes 0. If p' moves farther along the curve toward p, the average slope between p and p' decreases further, becoming negative, and drops to -1 when p' reaches p_3. If p' moves still closer to p the average slope now begins to increase: it becomes 0 again when p' reaches p_4; then keeps increasing and attains the value 1 when p' reaches p_5. And if p' moves farther along the curve toward p the same cycle is repeated: As p' approaches p and traverses a complete wave of the wave curve, the average slope between p and p' drops from 1 to -1, only to rise again from -1 to 1. Observe that if p' approaches p without limit, it must travel through infinitely many waves, since the curve as we have defined it generates this pattern. That is, as p' approaches p without limit the average slope between p and p' keeps oscillating between the values 1 and -1. Thus, as regards this slope, there can be no question of its limit nor of a definite slope of the curve at the point p. In other words the curve we have been considering has no tangent at p.

This relatively simple intuitable illustration demonstrates that a curve does not have to have a tangent at every point. It used to be thought, however, that intuition forced us to acknowledge that such a deficiency could occur only at isolated and exceptional points of a curve, never at all points. It was believed that a curve must possess an exact slope, or tangent, if not at every point, at least at an overwhelming majority of them. The mathematician and physicist Ampère, whose contribution to the theory of electricity is well known, attempted to prove this conclusion. His proof was false, and it was therefore a great surprise when Weierstrass announced the existence of a curve that lacked a precise slope or tangent at any point. Weierstrass invented the curve by an intricate and arduous calculation, which I shall not attempt to reproduce. But his result can today be achieved in a much simpler way, and this I shall attempt to explain, at least in outline.[7]

We start with the simple figure shown in Figure 4, which consists

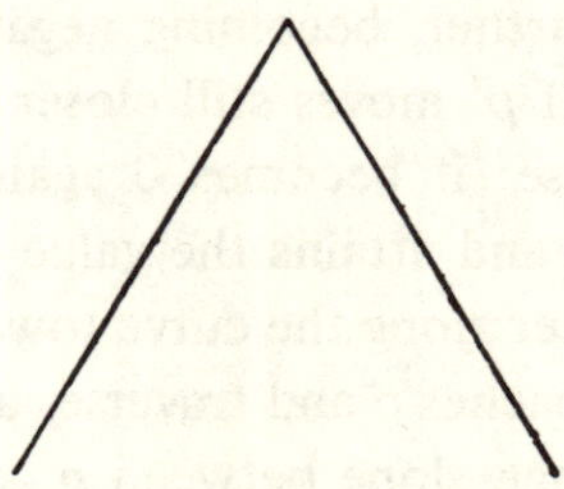

Fig. 4.

of an ascending and a descending line. The ascending line we shall replace, as shown in Figure 5, by a broken line of six parts, which first rises to half the height of the original line, then drops all the way down, then again rises to half height, continues on to full height, drops back again to half height, and finally rises once more to full height. Similarly we replace the descending line of Figure 4 by a broken line of six parts, which drops from full height to

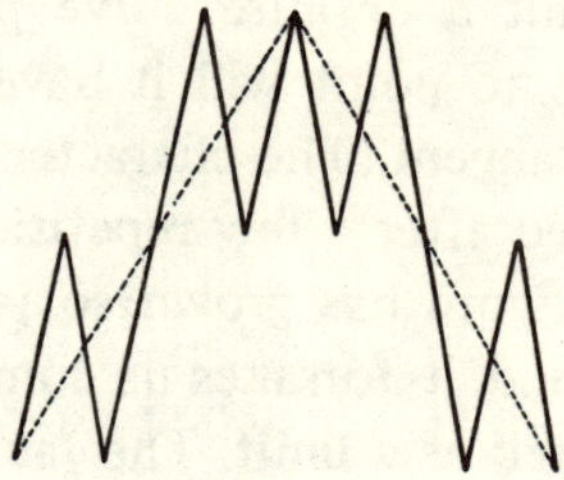

Fig. 5.

half height, rises again to full height, then again drops to half height and continues all the way down, rises once more to half height, and finally drops all the way down. From this figure composed of 12 line segments, we evolve by an analogous method the figure of 72 line segments, shown in Figure 6; that is by

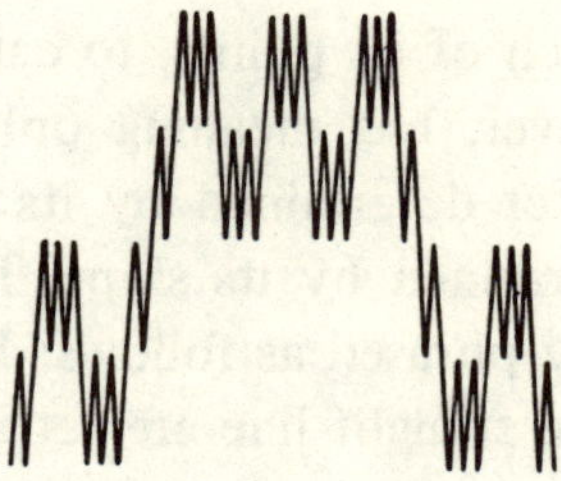

Fig. 6.

replacing every line segment of Figure 5 by a broken line of 6 parts. It is easy to see how this procedure can be repeated, and that it will lead to more and more complicated figures. There exists a rigorous proof (though I cannot give it here) that the succession of geometric objects constructed according to this rule approach without limit a definite curve possessing the desired property: namely, at no point will it have a precise slope, and hence at no point a tangent. The character of this curve entirely eludes intuition; indeed after a few repetitions of the segmenting process the evolving figure has grown so intricate that intuition can scarcely follow; and it forsakes us completely as regards the curve that is approached as a limit. The fact is that only thought, or logical analysis can pursue this strange object to its final form. Thus, had we relied on intuition in this instance, we should have remained in error, for intuition seems to force the conclusion that there cannot be curves lacking a tangent at any point.

This first example of the failure of intuition involves the fundamental concepts of differentiation; a second example can be derived from the fundamental concepts of integration. The basic problem of differentiation is: given the path of a moving point, to calculate its velocity, or given a curve, to calculate its slope; the basic problem of integration is the inverse: given the velocity of a moving point at every instant, to calculate is path, or given the slope of a curve at each of its points, to calculate the curve. This latter problem, however, has meaning only if the path of the moving point is in fact determined by its velocity, if the curve itself is actually determined by its slope. The question facing us can be more precisely phrased as follows: If two movable points whose track is a single straight line are set in motion in the same direction, at the same instant, from the same place on the line, and at every instant have the same velocity, must they remain together or can they become separated? – or: if two curves in a plane start from a common origin and continuously have the same slope, must they coincide in their entire course or can one of them rise above the other? The dictate of intuition is that the two moving points must always remain together, and that the two curves must coincide in their entire course; yet logical analysis shows that this is not necessarily so. The intuitive answer is true of course for ordinary curves and motions, but we can conceive of certain rather complicated motions and curves for which it is not true. I am sorry not to have the space to go into this in more detail; we must content ourselves with recognizing that in this instance also the apparent certainty of intuition proves to be deceptive.[8]

The foregoing examples of the failure of intuition involve the concepts of the calculus – a subject whose difficulty is acknowledged by the customary designation of 'higher mathematics'. Lest it be supposed, therefore, that intuition fails only in the more complex branches of mathematics, I propose to examine an occurrence of failure in the elementary branches.

At the very threshold of geometry lies the concept of the curve; everyone believes that he has an intuitively clear notion of what a curve is, and since ancient times it has been held that this idea could be expressed by the following definition: Curves are geometric figures generated by the motion of a point.[9] But, stay! In the year 1890 the Italian mathematician Giuseppe Peano (who is also renowned for his investigations in logic) proved that the geometric figures that can be generated by a moving point also include entire plane surfaces. For instance, it is possible to imagine a point moving in such a way that in a finite time it will pass through all the points of a square – and yet no one would consider the entire area of a square as simply a curve. With the aid of a few diagrams I shall attempt to give at least a general idea of how this space-filling motion is generated.[10]

Divide a square into four small squares of equal size (as shown in Figure 7) and join the centre points of these squares hy a continuous curve composed of line segments; now imagine a point moving in such a way that at uniform velocity it will traverse the curve in a finite time – say, in some particular unit of time. Next,

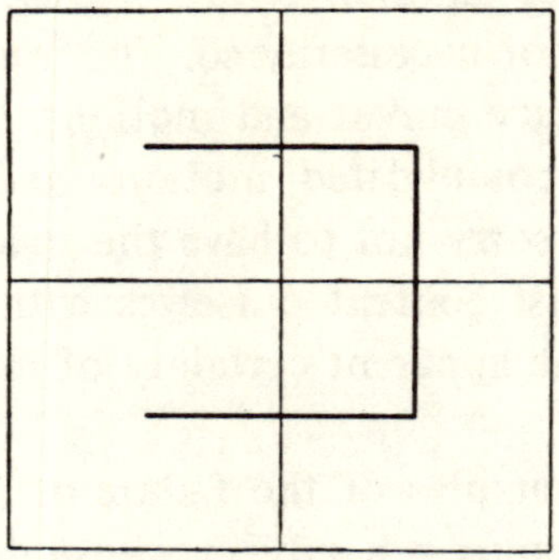

Fig. 7.

divide (Figure 8) each of the four small squares of Figure 7 into four smaller squares of equal size and connect the centre points of these 16 squares by a similar line, and again imagine the point moving so that in unit time it will traverse this second curve at

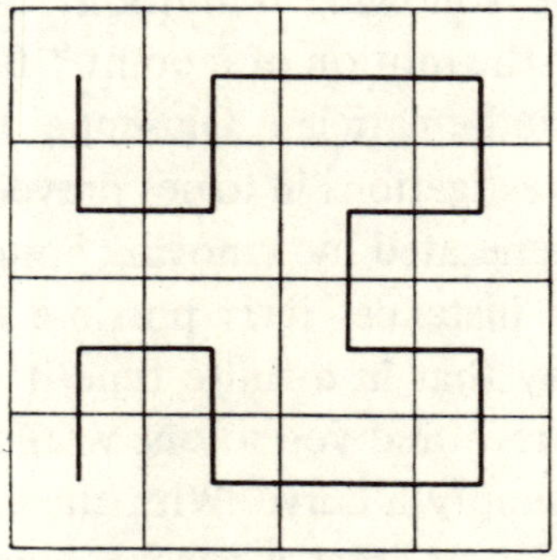

Fig. 8.

uniform velocity. Repeat this procedure (Figure 9) by dividing each of the 16 small squares of Figure 8 into four still smaller squares, connecting the centre points of these 64 squares by a third curve, and imagining the point to move so that in unit time

Fig. 9.

it will traverse this new system of lines at a uniform velocity. It is easy to see how this procedure is to be continued; Figure 10 shows one of the later stages, when the original square has been divided

Fig. 10.

into 4096 small squares. It is now possible to give a rigorous proof that the successive motions considered here approach without limit a definite course, or curve, that takes the moving point though all the points of the large square in unit time. This motion cannot possibly be grasped by intuition; it can only be understood by logical analysis.

While certain geometric objects which no one regards as curves (e.g., a square) can, contrary to intuition, be generated by the motion of a point, other objects that one would not hesitate to classify as curves cannot be so generated. Observe, for instance, the geometric shape shown in Figure 11. It is a wave curve which in the neighbourhood of the line segment *ab* (as shown in the figure)

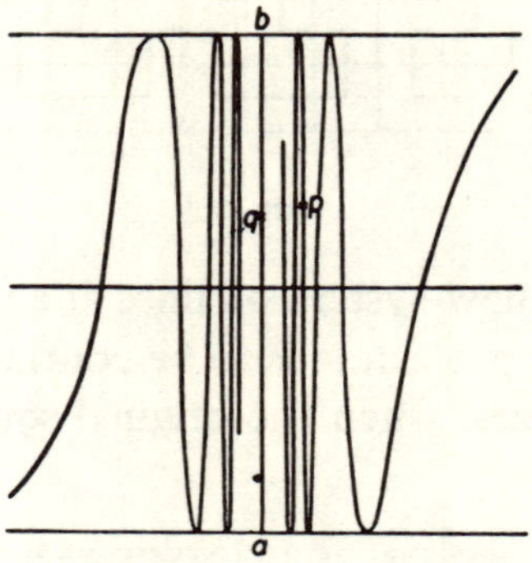

Fig. 11.

consists of infinitely many waves, whose lengths decrease without limit but whose amplitudes (in contrast to the curve of Figure 3), do not decrease. It is not difficult to prove that this figure, in spite of its linear character cannot be generated by the motion of a point; for no motion of a point is conceivable that would carry it through all the points of this wave curve in a finite time.

Two important questions now suggest themselves. 1. Since the time-honoured definition of a curve fails to cover the fundamental concept, what other more serviceable definition can be substituted for it? 2. Since the class of geometric objects that can be produced by the motion of a point does not coincide with the class of all curves, how shall the former class be defined? Today both questions are satisfactorily answered; I shall withhold for a moment the answer to the first question and speak briefly about the second.[11]

This was solved with the aid of a new geometric concept, 'connectivity in the small' [*Zusammenhang im Kleinen*] or 'local connectivity.'[12] Consider certain figures that can be generated by the motion of a point, such as (Figure 12) a line, a circle, or the area

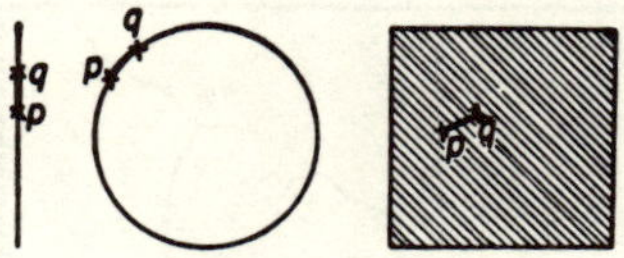

Fig. 12.

of a square. In each of these figures we fix our attention on two points p and q that lie very close together. It is evident that in each case we can move from p to q along a path that does not leave the confines of the figure and remains throughout in close proximity to p and q. This property is called (in an appropriately precise formulation) 'connectivity in the small'. The structure shown in Figure 11 does not have this property. Take for example the neighbouring points p and q; in order to move from p to q without leaving the curve it is necessary to traverse the infinitely many waves lying between them. This path does not remain in close proximity to p and q, since all the intervening waves have the same amplitude. Now it is important to realize that 'connectivity in the small' is the basic characteristic of figures that can be generated by the motion of a point. A line, a circle, and the area of a square can be generated by the motion of a point, because they are connected in the small; the construction in Figure 11 cannot be generated by the motion of a point, because it is not connected in the small.

We can convince ourselves by a second example of the undependability of intuition even as regards very elementary geometrical questions. Think of a map (Figure 13) showing the areas of three countries. On this map there will be boundary points at which two of the countries touch each other, but there may also be points at which all three countries come together – so called 'three-country corners', like the points *a* and *b* in Figure 13. Intuition

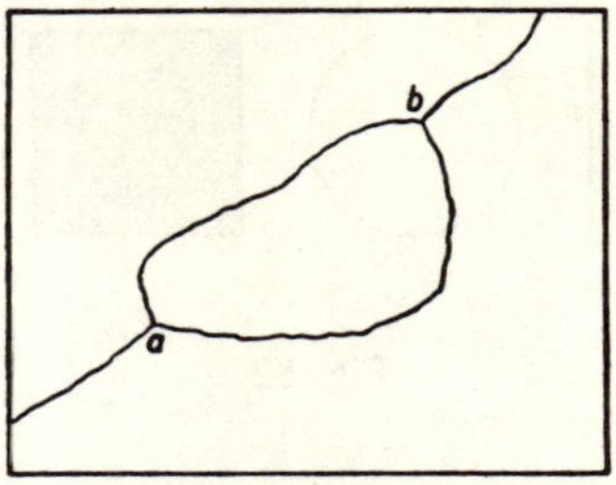

Fig. 13.

seems to indicate that these three-country corners can occur only at isolated points, that at the great majority of boundary points on the map only two countries will be in contact. Yet the Dutch mathematician L. E. J. Brouwer showed in 1910 how a map can be divided into three countries in such a way that at every boundary point all three countries will touch one another.[13] Let us attempt briefly to describe how this is done.

We start with the map shown in Figure 14, on which there are

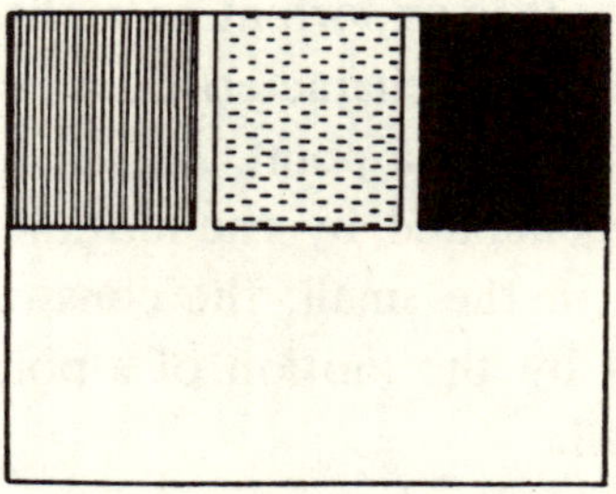

Fig. 14.

three different countries, one hatched (A), one dotted (B), and one solid (C); the unmarked remainder is unoccupied land. Country A, seeking to bring this land into its sphere of influence, decides to push through a corridor (Figure 15) which approaches within one

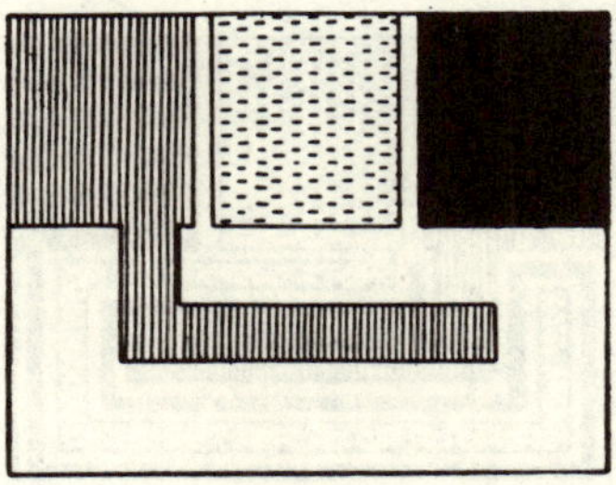

Fig. 15.

mile of every point of the unoccupied territory; but – in order to avoid trouble – the corridor is not to impinge upon either of the two other countries. After this has been done country B decides that it must make a similar move, and proceeds to drive a corridor into the remaining unoccupied territory (Figure 16) –

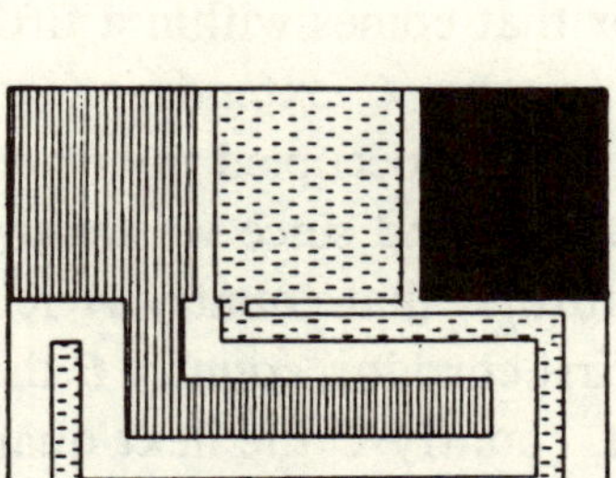

Fig. 16.

a corridor that comes within one-half mile of all the unoccupied points but does not touch either of the other two countries. Thereupon country *C*, deciding that it cannot lag behind, also extends a corridor (Figure 17) into the territory as yet unoccupied,

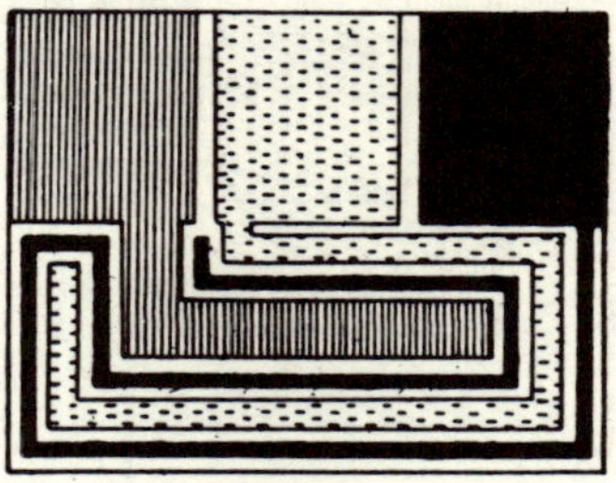

Fig. 17.

which comes to within a third of a mile of every point of this territory but does not touch the other countries. Country *A* now feels that it has been outwitted and proceeds to push a second corridor into the remaining unoccupied territory, which comes within a quarter of a mile of all points of this territory but does not touch the other two countries. The process continues: country *B* extends a corridor that comes within a fifth of a mile of every unoccupied point; country *C* extends one that comes within a sixth of a mile of every unoccupied point; country *A* starts over again, and so on and on. And since we are giving imagination free rein, let us assume further that country *A* required a year for the construction of its first corridor, country *B* the following half-year for its first corridor, country *C* the next quarter year for its first corridor; country *A* the next eighth of a year for its second, and so on; thus each succeeding extension is completed in half the time of its predecessor. It can easily be seen that after the passage of two years none of the originally unoccupied territory will remain unclaimed; moreover the entire map will then be divided among the three countries in such a fashion that at no point will only two of the countries touch each other, but instead all three countries will meet at every boundary point. Intuition cannot comprehend this pattern, although logical analysis requires us to accept it. Once more intuition has led us astray.

Because intuition turned out to be deceptive in so many instances, and because propositions that had been accounted true by intuition were repeatedly proved false by logic, mathematicians became more and more sceptical of the validity of intuition. They learned that it is unsafe to accept any mathematical proposition, much less to base any mathematical discipline on intuitive convictions. Thus a demand arose for the expulsion of intuition from mathematical reasoning, and for the complete formalization of mathematics. That is to say, every new mathematical concept was to be introduced through a purely logical definition; every mathematical proof was to be carried through by strictly logical means. The pioneers of this programme (to mention only the most famous) were Augustin Cauchy (1789–1857), Bernard Bolzano (1781–1848), Carl Weierstrass (1815–1897), Georg Cantor (1845–1918) and Richard Dedekind (1831–1916).

The task of completely formalizing mathematics, of reducing it entirely to logic, was arduous and difficult; it meant nothing less than a reform in root and branch. Propositions that had formerly been accepted as intuitively evident had to be painstakingly proved. To cite one example: the simple geometric proposition that "every closed polygon that does not cross itself divides the plane into two separate parts" requires a lengthy and highly complicated proof.[14] This is true to an even greater degree of the analogous proposition of solid geometry: "every closed polyhedron that does not intersect itself divides space into two separate parts."[15]

As the prototype of an *a priori* synthetic judgment based on pure intuition Kant cites the proposition that space is three-dimensional. But by present-day standards even this statement calls for searching logical analysis. First it is necessary to define purely logically what is meant by the 'dimensionality' of a geometric figure, or 'point set', and then it must be logically proved that the space of ordinary geometry – which is also the space of Newtonian physics – as embraced in this definition, is in fact three-dimensional. This proof was not achieved till recent times, in 1922, and then simultaneously by the Viennese mathematician K. Menger and the Russian mathematician P. Urysohn – the latter having since succumbed to a tragic accident at the height of his creative powers. I wish to give at least a sketchy explanation of how the dimensionality of a point set is defined.[16]

A point set is called null-dimensional if for each of its points there exists an arbitrarily small neighbourhood whose boundary contains no point of the set: for example, every set consisting of a finite number of points is null-dimensional (cf. Figure 18), but

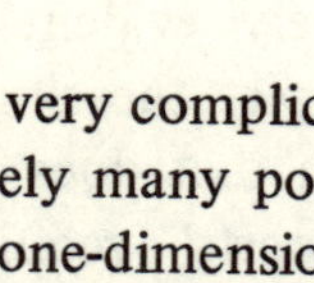

Fig. 18.

there are also a great many very complicated null-dimensional point sets that consist of infinitely many points. A point set that is not null-dimensional is called one-dimensional if for each of its points there is an arbitrarily small neighbourhood whose boundary has only a null-dimensional set in common with the point set. Every straight line, every figure composed of a finite number of straight lines, every circle, every ellipse, in short all geometrical constructs that we ordinarily designate as curves are one-dimensional in this sense (cf. Figure 19); in fact, even the geometric object shown in Figure 11, which we saw could not be generated by the motion of a point, is one-dimensional. Similarly, a point set that is neither

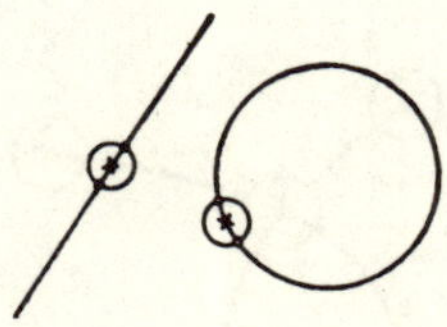

Fig. 19.

null-dimensional nor one-dimensional is called two-dimensional if for each of its points there is an arbitrarily small neighbourhood whose boundary has at the most a one-dimensional set in common with the point set. Every plane, every polygonic or circular plane area, every spherical surface, in short every geometric construct ordinarily classified as a surface is two-dimensional in this sense. A point set that is neither null-dimensional, one-dimensional, nor two-dimensional is called three-dimensional if for each of its points there is an arbitrarily small neighbourhood whose boundary has at most a two-dimensional set in common with the point set. It can be proved – not at all simply, however – that the space of ordinary geometry is a three-dimensional point set in the foregoing theory.

This theory provides what we have been seeking, a fully satisfactory definition of the concept of a curve.[17] The essential characteristic of a curve turns out to be its one-dimensionality. But beyond that the theory also makes possible an unusually precise and subtle analysis of the structure of curves, about which I should like to comment briefly. A point on a curve is called an end point if there are arbitrarily small neighbourhoods surrounding it, each of whose boundaries has only a single point in common with the curve (cf. points *a* and *b* in Figure 20). A point on the curve that is not an end point is called an ordinary point if it has arbitrarily small neighbourhoods each of whose boundaries has

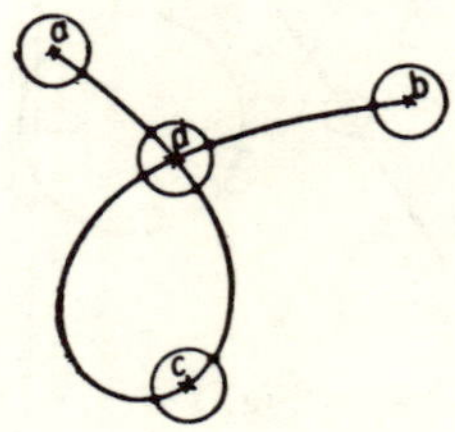

Fig. 20.

exactly two points in common with the curve (cf. point *c* in Figure 20). A point on a curve is called a branch point if the boundary of any of its arbitrarily small neighbourhoods has more than two points in common with the curve (cf. point *d* in Figure 20).

Intuition seems to indicate that the end points and branch points of a curve are exceptional cases; that they can occur only sporadically, that it is impossible for a curve to be made up of nothing but end points or of branch points. This intuitive conviction is specifically confirmed by logical analysis as far as end points are concerned; but as regards branch points it has been refuted. The Polish mathematician W. Sierpiński proved in 1915 that there are curves *all of whose points are branch points.* Let us attempt to visualize how this comes about.

Suppose that an equilateral triangle has been inscribed within another equilateral triangle (as shown in Figure 21) and the interior of the inscribed triangle erased (hatched in Figure 21); there remain three equilateral triangles with their sides. In each of these

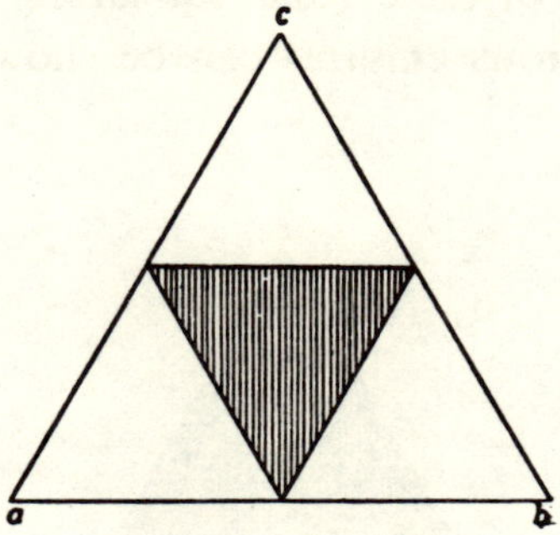

Fig. 21.

three triangles (Figure 22) inscribe an equilateral triangle and again erase its interior; there are now left nine equilateral triangles together with their sides. In each of these nine triangles an equilateral triangle is to be inscribed and its interior erased so that 27 equilateral triangles are left. Imagine this process continued

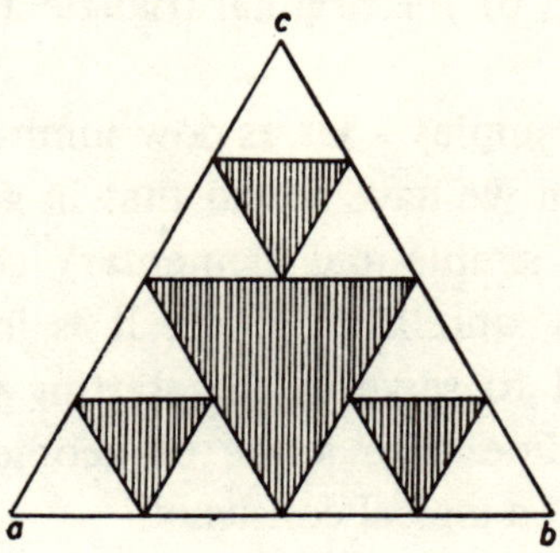

Fig. 22.

indefinitely. (Figure 23 shows the fifth step, where 243 triangles remain.) The points of the original equilateral triangle that survive the infinitely numerous erasures can be shown to form a curve,

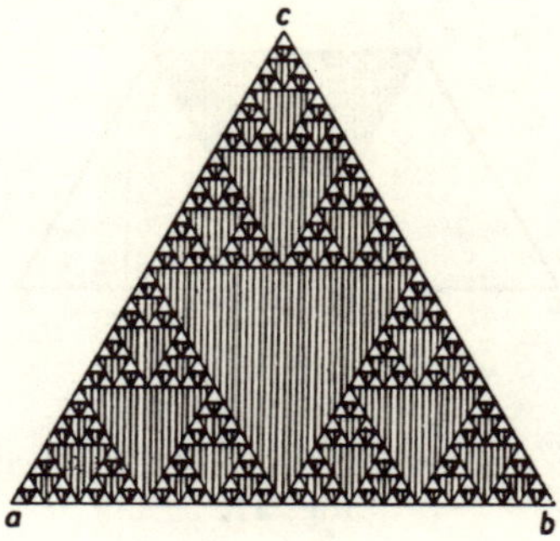

Fig. 23.

and specifically a curve all of whose points, with the exception of the vertex points *a, b, c*, of the original triangle, are branch points. From this it is easy to obtain a curve with all its points branch points; for instance, by distorting the entire figure so that the three vertices *a, b, c*, of the original triangle are brought together in a single point.

But enough of examples – let us now summarize what has been said. Again and again we have found that in geometric questions, and indeed even in simple and elementary geometric questions, intuition is a wholly unreliable guide. It is impossible to permit so unreliable an aid to serve as the starting point or basis of a mathematical discipline. The space of geometry is not a form of pure intuition, but a logical construct.

The way is then open for other non-contradictory logical constructs in the form of spaces differing from the space of ordinary geometry; space, for instance, in which the so-called

Euclidean parallel postulate is replaced by a contrary postulate (non-Euclidean spaces), spaces whose dimensionality is greater than three, non-Archimedean spaces. I shall say a few words about the last named.

The possibility of measuring the length of a line segment by a real number, and the possibility that follows from this, namely, fixing the position of a point, as is done in analytic geometry, by assigning real numbers as its 'coordinates', rests on the so-called 'postulate of Archimedes'.[18] This postulate reads as follows: given two lengths, there is always a multiple of the first that is greater than the second. As logical constructs, however, spaces can be devised in which the Archimedean postulate is replaced by its opposite, that is, in which there are lengths that are greater than any multiple of a given length.[19] Hence in these spaces infinitely large and infinitely small lengths can exist (as determined by any arbitrarily chosen unit of measure); while in the space of ordinary geometry there are no infinitely large and infinitely small lengths. In a 'non-Archimedean' space, lengths can be measured, and a system of analytical geometry can be developed. Of course, the real numbers of ordinary arithmetic are of no help in this geometry, but one uses 'non-Archimedean' number systems, which can be interpreted and applied in calculation as well as the real numbers of ordinary arithmetic.[20]

But what are we to say to the often heard objection that the multi-dimensional, non-Euclidean, non-Archimedean geometries, though consistent as logical constructs, are useless in arranging our experience because they are non-intuitional? For projecting our experience, it is said, only the conventional three-dimensional, Euclidean, Archimedean geometry is usable, for it is the only one that satisfies intuition. My first comment on this score – and this is the point of my entire lecture – is that ordinary geometry itself by no means constitutes a supreme example of the intuitive process. The fact is that *every* geometry – three-dimensional as well as multi-dimensional, Euclidean as well as non-Euclidean, Archimedean as well as non-Archimedean – is a logical construct. Traditional physics is responsible for the fact that until recently the logical construction of three-dimensional, Euclidean, Archimedean space has been used exclusively for the ordering of our experience. For several centuries, almost up to the present day, it served this purpose admirably; thus we grew used to operating with it. This habituation to the use of ordinary geometry for the ordering of our experience explains why we regard this geometry as intuitive; and every departure from it unintuitive, contrary to intuition, intuitively impossible. But as we have seen, such 'intuitional impossibilities', also occur in ordinary geometry. They appear as soon as we no longer restrict ourselves to the geometrical entities with which we have long been familiar, but instead reflect upon objects that we had not thought about before.

Modern physics now makes it appear appropriate to avail ourselves of the logical constructs of multi-dimensional and non-Euclidean geometries for the ordering of our experience. (Although we have as yet no indication that the inclusion of non-Archimedean geometry might prove useful, this possibility is by no means excluded.) But these advances in physics are so recent that we are not yet accustomed to the manipulation of these logical constructs; hence they still seem an affront to intuition.

The theory that the earth is a sphere was also once an affront to intuition. This hypothesis was widely rejected on the grounds that the existence of the antipodes was contrary to intuition. However, we have got used to the idea and today it no longer occurs to anyone to pronounce it impossible because it conflicts with intuition.

Physical concepts are also logical constructs and here too we can see clearly how concepts whose application is familiar to us acquire an intuitive status which is denied to those whose application is unfamiliar. The concept 'weight' is so much a part of common experience that almost everyone regards it as intuitive. The concept 'moment of inertia', however, does not enter into most persons' activities and is therefore not regarded by them as intuitive. Yet among experimental physicists and engineers, who constantly work with it, 'moment of inertia' possesses an intuitive status equal to that generally accorded to 'weight'. Similarly, the concept 'potential difference' is intuitive for the electrical technician, but not for most men.

If the use of multi-dimensional and non-Euclidean geometries for the ordering of our experience continues to prove itself so that we become more and more accustomed to dealing with these logical constructs; if they penetrate into the curriculum of the schools; if we, so to speak, learn them at our mother's knee, as we now learn three-dimensional Euclidean geometry – then nobody will think of saying that these geometries are contrary to intuition. They will be considered as deserving of intuitive status as three-dimensional Euclidean geometry is today. For it is not true, as Kant urged, that intuition is a pure *a priori* means of knowledge, but rather that it is force of habit rooted in psychological inertia.

NOTES

1 H. Mark, *Die Erschütterung der klassischen Physik durch das Experiment* and H. Thirring, *Die Wandlung des Begriffssystemes der Physik.*

2 For further details see W. Heisenberg, *Die physikalischen Prinzipien der Quantentheorie* (Leipzig, 1930) [E. T. *The Physical Principles of the Quantum Theory* (Chicago, 1930)] and Thirring's lecture quoted in Note 1.

3 K. Menger, *Die neue Logik* [E. T. 'The New Logic' as Chapter I of K. Menger, *Selected Papers in Logic and Foundations, Didactics, Economics* in this Collection].

4 The main work here is Whitehead–Russell, *Principia Mathematica* (Cambridge, [1]1910–3, [2]1925). Presentation for the general reader in B. Russell, *Introduction to Mathematical Philosophy* (London, 1919) [G. T. Munich, 1923].

5 The considerations about 'velocity' and 'slope' that follow will be found in greater detail in Hahn–Tietze, *Einführung in die Elemente der höheren Mathematik* (Leipzig, 1925) pp. 153ff., 190ff.

6 [In Austria, at that time.]

7 See H. Hahn, *Jahresber. d. Deutschen Math.-Vereinigung* 26 (1918) p. 281, and L. Bieberbach, *Differential- u. Integralrechrung* I (Leipzig, 1917) p. 104, for a precise mathematical treatment of what follows.

8 The reader may turn to H. Hahn, *Monatshefte f. Math. u. Phys.,* **16** (1905) p. 161, which deals with motions (or curves) that assume infinite velocities (or slopes of infinite value).

9 By 'motion' is here meant a change of place that proceeds continuously, i.e. one in which between two instants sufficiently close to one another the point in motion passes through arbitrarily few positions. Such a point will make no 'jumps'.

10 Detailed discussion in H. Hahn, *Theorie der reellen Funktionen* (Berlin, 1921) p. 146, K. Menger, *Kurventheorie* (Leipzig, 1932) p. 10.

11 This question was answered in 1913–14 by H. Hahn and S. Mazurkiewicz. Later accounts in F. Hausdorff, *Mengenlehre*^{2} (Berlin, 1927) p. 205; H. Hahn, *Reelle Funktionen* (Leipzig, 1932) p. 164; K. Menger, *Kurventheorie* (Leipzig, 1932) p. 31.

12 [The German expression, not easy to translate, may also be rendered as 'connection . . . ' and 'connectedness in the small', and the function describing the property is sometimes given as a 'piecewise continuous function'. Translator.]

13 L. E. J. Brouwer, *Math. Annalen* **68** (1910) p. 427. In the present discussion we make use of an intuitive interpretation suggested by the Japanese mathematician Wada. A point is called a 'boundary point' if in each of its neighbourhoods lie points of various countries. Three countries meet in a boundary point if in each of its neighourhoods points are to be found of each of the three countries.

14 A proof by H. Hahn in *Monatsh. f. Math. u. Phys.* **19** (1908) p. 289.

15 Lilly Hahn, *Monatsh. f. Math. u. Phys.* **25** (1914) p. 303; N. J. Lennes, *American Journal of Mathematics* **33** (1911) p. 37.

16 Detailed presentation in K. Menger, *Dimensionstheorie* (Leipzig, 1928).

17 Detailed presentation in K. Menger, *Kurventheorie* (Leipzig, 1932).

18 It should properly be called the postulate of Eudoxus. Eudoxus' dates are 408–355 B.C.; Archimedes', 287–212 B.C.

19 The first to investigate thoroughly the properties of non-Archimedean spaces was the Italian mathematician G. Veronese, *Grundzüge der Geometrie von mehreren Dimensionen und mehreren Arten gradliniger Einheiten* (Leipzig, 1894).

20 On non-Archimedean number systems see H. Hahn, *S.-B. Wien* (Math.-Nat. Wiss. K1.) **116** (1907) p. 601.

DOES THE INFINITE EXIST?*

Since antiquity the mind of man has continually been vexed by the question of the infinite. Again and again it has been emphatically denied that anything infinite could exist or that any infinity could be conceived by man: again and again thinkers have arisen who found this perfectly possible. Aristotle had taught that something complete and infinite was in no way possible. The philosophers of the Middle Ages, dominated on the one hand by the authority of Aristotle (whom they styled 'the Philosopher' without further qualification) but on the other hand and to at least the same extent by the authority of the Church, never tired of discussing the question how Aristotle's doctrine of the impossibility of the infinite could be reconciled with the Church's doctrine of the omnipotence of God. St. Thomas Aquinas, refining and rendering more precise Aristotle's thesis taught that nothing infinite could be given, whereas not a few of the impressive succession of Schoolmen – particularly the nominalists – defended the opposite thesis.[1] A partial result, at all events, was the development of an admirable degree of logical rigour, which was completely lost in succeeding centuries and in many points only regained in the critical mathematics of the nineteenth century. Of modern thinkers we will here mention only some of the greatest and only those who had mathematical interests. Descartes expressly declines to discuss the infinite in any way: "We shall not weary ourselves", he writes in his *Principia,*[2] "with disputes about the infinite, since, finite as we are, it would be perverse to attempt to make any

* First published in *Alte Probleme – Neue Lösungen in den exakten Wissenschaften*, Fünf Wiener Vorträge, Zweiter Zyklus, Leipzig and Vienna, 1934.

determinations about it, turning it, as it were, into something finite that we could conceive . . . since only one who considers his own mind infinite can feel obliged to reflect upon this subject." Leibniz, on the other hand, writes in a letter[3]: "I am so much in favour of the actual infinite that, so far from admitting, as is commonly said, that nature abhors it, I claim that nature everywhere shows a predilection for it, the better to mark the perfection of its own Author. Hence I believe that every part of matter is not merely divisible but actually divided and that consequently the smallest of particles ought to be regarded as a world full of an infinity of different creatures." But he also teaches, agreeing for once with his polar opposite Locke, "We have no idea of an infinite space, and nothing is more palpable than the absurdity of an actual idea of an infinite number." In 1831 C. F. Gauss, who is honoured as the greatest of all mathematicians, wrote the often-quoted words,[4] "thus I protest against the use of an infinite quantity as something completed, which is never allowed in mathematics", and the same was the opinion of the great mathematician A. Cauchy, whose thought did much to determine the course of nineteenth-century mathematics. Quite other was the opinion of his contemporary, the Austrian B. Bolzano, who shares with Cauchy the credit for laying the foundations that provided analysis with an exact and critical basis. His fate also was typically Austrian, in that he was little regarded by his contemporaries, so that his influence on subsequent developments cannot remotely be compared with that of Cauchy. In 1847–8 he wrote a small work *Paradoxes of the Infinite,*[5] first printed after his death, in which he attempted "to see the appearance of contradiction attaching to these mathematical paradoxes as what it is, as *mere* appearance" and in this way to make the infinite an object of scientific investigation. It was, however, Georg Cantor who achieved decisive success in this direction, when in the years 1871–84 he created a completely new and very special mathematical discipline, the theory of sets, in which for the first time in a thousand years of argument back and forth, a theory of infinity with all the incisiveness of modern mathematics, was given its foundations.[6]

Like so many other new creations this one began with a very simple idea. Cantor asked himself: "What do we mean when we say of two finite sets that they consist of equally many things, that they have the same number, that they are equivalent?" Obviously nothing more than this, that between the members of the first set and those of the second a correspondence can be effected by which each member of the first set matches exactly a member of the second set, and likewise each member of the second set matches one of the first. A correspondence of this kind is called 'reciprocally unique', or simply 'one-to-one'. The set of the fingers of the right hand is equivalent to the set of fingers of the left hand, since between the fingers of the right hand and those of the left hand a one-to-one pairing is possible. Such a correspondence is obtained, for instance, when we place the thumb on the thumb, the index finger on the index finger, and so on. But the set of both ears and the set of the fingers of one hand are not equivalent, since in this instance a one-to-one correspondence is obviously impossible; for if we attempt to place the fingers of one hand in correspondence with our ears, no matter how we contrive there will necessarily be some fingers left over to which no ears correspond. Now the number (or cardinal number) of a set is obviously a characteristic that it has in common with all equivalent sets, and by which it distinguishes itself from every set not equivalent to itself. The number 5, for instance, is the characteristic which all sets equivalent to the set of the fingers of one hand have in common, and which distinguishes them from all other sets.

Thus we have the following definitions. Two sets are called equivalent if between their respective members a one-to-one correspondence is possible; and the characteristic that one set has in common with all equivalent sets, and by which it distinguishes itself from all other sets not equivalent to itself is called the (cardinal) number of that set. And now we make the fundamental assertion that in these definitions the finiteness of the sets considered is in no sense involved; the definitions can be applied as readily to infinite sets as to finite sets. The concepts 'equivalent' and 'cardinal number' are thereby transferred to sets of infinitely many objects. The cardinal numbers of finite sets, i.e., the numbers 1, 2, 3, . . . , are called natural numbers; the cardinal numbers of infinite sets Cantor calls 'transfinite cardinal numbers' or 'transfinite powers'.

But are there really any infinite sets? We can convince ourselves of this at once by a very simple example. There are obviously infinitely many different natural numbers; hence the set of all the natural numbers contains infinitely many members; it is an infinite set. Now then, those sets that are equivalent to the set of all natural numbers, whose members can be paired in one-to-one correspondence with the natural numbers, are called *denumerably infinite sets*. The meaning of this designation can be explained as follows. A set with the cardinal number 'five' is a set whose members can be put in one-to-one correspondence with the first five natural numbers, that is, a set that can be numbered by the integers 1, 2, 3, 4, 5, or counted off with the aid of the first five natural numbers. A denumerably infinite set is a set whose members can be put in one-to-one correspondence with the totality of natural numbers. According to our definitions all denumerably infinite sets have the same cardinal number; this cardinal number must now be given a name, just as the cardinal number of the set of the fingers on one hand was earlier given the name 5. Cantor gave this cardinal number the name 'aleph-null', written $\aleph_0$ (Why he gave it this rather bizarre name will become clear later.) The number $\aleph_0$ is thus the first example of a transfinite cardinal number. Just as the statement "a set has the cardinal number 5" means that its members can be put in one-to-one correspondence with the fingers of the right hand, or – what amounts to the same thing – with the integers 1, 2, 3, 4, 5, so the statement "a set has the cardinal number $\aleph_0$" means that its elements can be put in one-to-one correspondence with the totality of natural numbers.

If we look about us for examples of denumerably infinite sets we arrive immediately at some highly surprising results. The set of all natural numbers is itself denumerably infinite: this is self-evident, for it was from this set that we defined the concept 'denumerably infinite'. But the set of all even numbers is also denumerably infinite, and has the same cardinal number $\aleph_0$ as the set of all natural numbers, though we would be inclined to think that there are far fewer even numbers than natural numbers. To prove this proposition we have only to look at the correspondence diagrammed in Figure 1, that is, to put each natural number opposite its double. It may clearly be seen that there is a one-to-one correspondence between all natural and all even numbers, and thereby our point is established. In exactly the same way it can be shown that the set of all odd numbers is denumerably infinite. Even more surprising is the fact that the set of *all pairs of natural numbers* is denumerably infinite. In order to understand this we have merely to arrange the set of all pairs of natural number 'diagonally' as indicated in Figure 2, whereupon we at once obtain the one-to-one correspondence shown in Figure 3 between all natural numbers and all pairs of natural numbers. From this follows the conclusion, which Cantor discovered while still a

1	2	3	4	5	6	. . .
↕	↕	↕	↕	↕	↕	
2	4	6	8	10	12	. . .

Fig. 1.

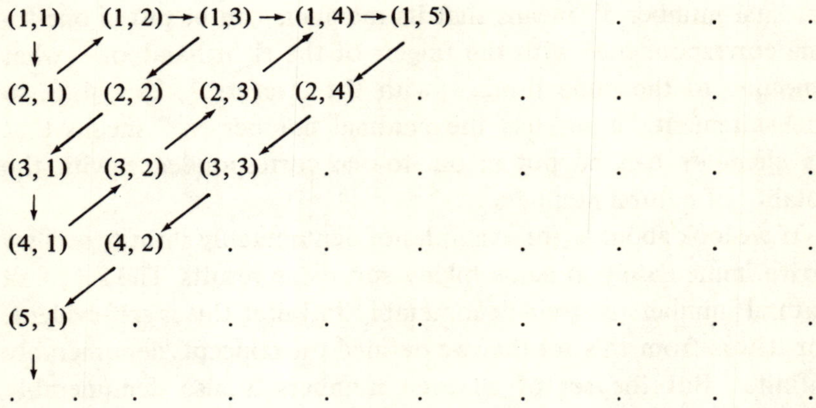

Fig. 2.

1	2	3	4	5	6	7	8	9
↕	↕	↕	↕	↕	↕	↕	↕	↕
(1, 1)	(2, 1)	(1, 2)	(1, 3)	(2, 2)	(3, 1)	(4, 1)	(3, 2)	(2, 3)
10	11	12	13	14	15	.	.	.
(1, 4)	(1, 5)	(2, 4)	(3, 3)	(4, 2)	(5, 1)	.	.	.

Fig. 3.

student, that the set of all rational fractions (i.e., the quotients of two whole numbers, like 1/2, 2/3, etc.) is also denumerably infinite, or equivalent to the set of all natural numbers, though again one might suppose that there are many, many more fractions than there are natural numbers. What is more, Cantor was able to prove that the set of all so-called algebraic numbers, that is, the set of all numbers that satisfy an algebraic equation of the form

$$a_0x^n + a_1x^{n-1} + \ldots + a_{n-1}x + a_n = 0$$

with integral coefficients $a_0, a_1, \ldots a_n$, is denumerably infinite.

At this point, the reader may ask whether, in the last analysis, *all* infinite sets are not denumerably infinite – that is, equivalent? If this were so, we should be sadly disappointed; for then, alongside the finite sets there would simply be infinite ones which would all be equivalent, and there would be nothing more to say about the matter. But in the year 1874 Cantor succeeded in proving that there are also infinite sets that are not denumerable; that is to say, there are other infinite numbers, transfinite cardinal numbers differing from aleph-null. Specifically, Cantor proved that the set of all so-called real numbers (i.e., composed of all whole numbers, plus all fractions, plus all irrational numbers) is non-denumerably infinite. The proof is so simple that I can give his reasoning here. It is sufficient to show that the set of all real numbers between 0 and 1 is not denumerably infinite, for then the set of *all* real numbers obviously cannot be denumerably infinite. In the proof we make use of the familiar fact that every real number between 0 and 1 can be expressed as an infinite decimal fraction. The statement that is to be proved, that the set of all real numbers between 0 and 1 is not denumerably infinite, can also be phrased as follows: "A *denumerably* infinite set of real numbers between 0 and 1 cannot be the set of *all* real numbers between 0 and 1", or thus: "Given a denumerably infinite set of real numbers between 0 and 1, there will always be a real number between 0 and 1 that does not belong to the given set." To prove this let us imagine a denumerably infinite set of real numbers between 0 and 1 as given; then, since this set is denumerably infinite, the real numbers that occur in it can be put in one-to-one correspondence with the natural numbers. Let us write down as a continuing decimal fraction the real number that in this arrangement corresponds to the natural number 1, and under it the real number that corresponds to the natural number 2, under it again the real number that corresponds to the natural number 3, and so on. We would get something like this:

0.20745 . . .
0.16238 . . .
0.97126 . . .
. . . .

Now one can, in fact, at once write down a real number between 0 and 1 which does *not* occur in the given denumerably infinite set of real numbers between 0 and 1. Take as its first digit one differing from the first digit of the decimal in the first row, say 3; as its second digit one differing from the second digit of the decimal in the second row, say 2; as its third digit one differing from the third digit of the decimal in the third row, say 5; and so on. It is clear that by this procedure we obtain a real number between 0 and 1 which differs from all the given infinitely many real numbers of our set, and this is precisely what we sought to prove was possible.

It has thus been shown that the set of natural numbers and the set of real numbers are not equivalent; that these two sets have different cardinal numbers. The cardinal number of the set of real numbers Cantor called the 'power of the continuum'; we shall designate it by c. Earlier it was noted that the set of all algebraic numbers is denumerably infinite, and we just now saw that the set of all real numbers is not denumerably infinite, hence there must be real numbers that are not algebraic. These are the so-called transcendental numbers, whose existence is demonstrated in the simplest way conceivable by Cantor's brilliant train of reasoning.

It is well known that the real numbers can be put in one-to-one correspondence with the points of a straight line; hence c is also the cardinal number of the set of all points of a straight line. Surprisingly Cantor was also able to prove that a one-to-one pairing is possible between the set of all points of a plane and the set of all points of a straight line. These two sets are thus equivalent, that is to say, c is also the cardinal number of the set of all points of a plane, though here too we should have thought that a plane would contain a great many more points than a straight line. In fact, as Cantor has shown, c is the cardinal number of *all* points of three-dimensional space, or even of a space of any number of dimensions.

We have discovered two different transfinite cardinal numbers $\aleph_0$ and c, the power of the denumerably infinite sets and the power of the continuum. Are there yet others? Yes, there certainly are infinitely many different transfinite cardinal numbers; for given any set M, a set with a higher cardinal number can at once be indicated, since the set of all possible subsets of M has a higher cardinal number than the set M itself. Take, for example, a set of three elements, such as the set of the three figures 1, 2, 3. Its partial sets are the following: 1; 2; 3; 1, 2; 2, 3; 1, 3; thus the number of the partial sets is more than three. Cantor has shown that this is generally true,[7] even for infinite sets. For example, the set of all possible point-sets of a straight line has a higher cardinal number than the set of all points of the straight line, that is to say, its cardinal number is greater than c.

What is now desired is a general view of all possible transfinite cardinal numbers. As regards the cardinal numbers of finite sets, the natural numbers, the following simple situation prevails: Among such sets there is one that is the smallest, namely 1; and if a finite set M with the cardinal number m is given, a set with the next-larger cardinal number can be formed by adding one more object to the set M. What is the rule in this respect with regard to infinite sets? It can be shown without difficulty that among the transfinite cardinal numbers, as well as among the finite ones, there is one that is the smallest, namely $\aleph_0$, the power of denumerably infinite sets (though we must not think that this is self-evident, for among all positive fractions, for instance, there is none that is the smallest). It is, however, not so easy as it was in the case of finite sets to form the next-larger to a transfinite cardinal number; for whenever we add one more member to an infinite set we do not get a set of greater cardinality, only one of equal cardinality. But Cantor also solved this difficulty, by showing that there is a next-larger to every transfinite cardinal number (which again is by no means self-evident: there is no next-larger to a fraction, for instance) and by showing how it is obtained. We cannot go into his proof here, since this would take us too far into the realm of pure mathematics. It is enough for us to recognize the fact that there is a smallest transfinite cardinal number, namely $\aleph_0$; after this there is a next-larger, which is called $\aleph_1$; after this there is again a next-larger, which is $\aleph_2$; and so on. But this still does not exhaust the class of transfinite cardinals; for if it be assumed that we have formed the cardinal numbers $\aleph_0, \aleph_1, \aleph_2, \ldots \aleph_{10}, \ldots \aleph_{100}, \ldots \aleph_{1000}, \ldots$ that is, all alephs ($\aleph_n$) whose index n is a natural number, then there is again a first transfinite cardinal number larger than any of these – Cantor called it $\aleph_\omega$ – and a next-larger successor $\aleph_{\omega+1}$, and so on and on.

The successive alephs formed in this manner represent all possible transfinite cardinal numbers, and hence the power c of the continuum must occur among them. The question is which aleph is the power of the continuum? This is the famous problem of the continuum. We already know that it cannot be $\aleph_0$, since the set of all real numbers is non-denumerably infinite, that is to say, not equivalent to the set of natural numbers. Cantor took $\aleph_1$ to be the power of the continuum. The question, however, remains open, and for the present we see no trace of a path to its solution.[8]

I must here call attention to a detail of logic that was first noticed by Ernst Zermelo some time after the formulation of the theory of sets by Cantor. In the proof that $\aleph_0$ is the smallest transfinite cardinal number, as well as in the proof that every transfinite cardinal number occurs in the aleph series, one makes an assumption that is also used in other mathematical proofs without its being explicitly recognized. This assumption, which, as it seems, cannot be derived from the other principles of logic, is known as the postulate, or the axiom of choice,[9] and may be stated as follows: Given a set of M sets, no two of which have a common member, then there exists a set that has exactly one member in common with each of the sets M. The postulate raises no difficulty when dealing with a finite collection of sets, for one can select a member from each of these sets in a finite number of operations. When one member has been selected from each of the given sets, our task is done for we then have formed a new set that has exactly one member in common with each of the given sets. Neither is there any difficulty if infinitely many sets are given and at the same time a rule is provided that distinguishes one member of each of these sets. The set composed of the objects denoted by the rule is the one we seek, since it also has exactly one member in common with each of the given sets. But if infinitely many sets are given and no rule is formulated that distinguishes one member in each of them, we cannot proceed as we did in the case of a finite number of sets. For if we started to select arbitrarily one member from each of the given sets, we could of course never complete the task and thus never could obtain a set having one member in common with each of the given sets. Hence the assertion that such a set exists in every case represents a special logical postulate. The famous English logician and philosopher, Bertrand Russell, has made this clear in an ingenious and amusing illustration. In civilized countries shoes are so designed that the right and left shoe of each pair can readily be distinguished; but the distinction between right and left cannot be made for pairs of stockings. Let us imagine an infinitely rich man (a millionaire or a billionaire will not do, he must be an 'infinitillionaire') who is infinitely eccentric and owns a collection of infinitely many pairs of shoes and infinitely many pairs of stockings. This fortunate man, it may be observed, has at his disposal a set of shoes that contains exactly one shoe from each pair; for instance, the infinite set composed of the right shoes, one from each pair. But how is he to obtain a set that contains exactly one stocking from each pair of stockings? This is, of course, only a facetious illustration, yet the principle itself represents a serious and important logical discovery, which bears on many problems. For the moment, however, let us leave it; we shall return to it.

On the basis of this rather sketchy description of the structure of the theory of sets the answer to the question "Is there an infinity?" appears to be an unqualified "Yes". There are not only, as Leibniz had already asserted, infinite sets, but there are even what Leibniz had denied, infinite numbers, and it can also be shown that it is quite possible to operate with them, in a manner similar to, if not identical with that used for finite natural numbers.

But now we must took with a critical eye at what has been accomplished. When existence is asserted of some entity in mathematics by the words 'there is' or 'there are' and non-existence by the words 'there is/are not', evidently some thing entirely different is meant than when these expressions are used in everyday life, or in geography, say, or natural history. Let us make this clear by a few examples. Since about the time of Plato it has been known that there are regular bodies with four sides (tetrahedrons), with six sides (hexahedrons or cubes), with eight sides (octahedrons), with twelve sides (dodecahedrons), and with twenty sides (icosahedrons). There are no regular bodies other than these five. Now it is obvious, that the expressions 'there are' and 'there are no' have a different meaning as used in the two preceding sentences about mathematics than in a geographical statement, such as:

there are mountains over 25,000 feet high, but there are no mountains 35,000 feet high. For even though mathematics teaches that there are cubes and icosahedrons, yet in the sense of physical existence, there are no cubes and no icosahedrons. The most beautiful rock-salt crystal is not an exact mathematical cube, and a model of an icosahedron, however well constructed, is not an icosahedron in the mathematical sense. While it is fairly clear what is meant by the expressions 'there is' or 'there are' as used in the sciences dealing with the physical world, it is not at all clear what mathematics means by such existence statements. On this point indeed there is no agreement whatever among scholars, whether they be mathematicians or philosophers. So many different interpretations are represented among them that one might almost say: *quot capita, tot sententiae* ("as many meanings as individuals"). However, if we stick to essentials we can perhaps distinguish three basically different points of view on this subject. These I shall describe briefly.

The first can be designated the realistic or the Platonic position. It ascribes to the objects of mathematics a real existence in the world of ideas; the physical world we may note is merely an imperfect image of the world of ideas. Thus, there are no perfect cubes in the physical world, only in the world of ideas; through our senses we can comprehend only the physical world, but it is in thought that we comprehend the world of ideas. A mathematical concept derives its existence from the real object corresponding to it in the world of ideas; a mathematical statement is true if it correctly represents the real relationship of the corresponding objects in the world of ideas. The second view which we can call the intuitionistic or the Kantian position, is to the effect that we possess pure intuition; mathematics is construction by pure intuition, and a mathematical concept 'exists' if it is constructible by pure intuition. The philosophical formulation of this idea might be popularly expressed somewhat as follows: "If there is indeed no perfect cube in the physical world I can at least imagine a perfect cube." The third view is best called the logistic position. If one sought to trace its historical roots one could perhaps connect it with the nominalist school of scholastic philosophy; thus it may be called the nominalist position. According to this view mathematics is a purely logical discipline and, like logic, is carried on entirely within the confines of language; it has nothing whatever to do with reality, or with pure intuition; on the contrary it deals exclusively with the use of signs or symbols. These signs or symbols can be used as we like, in conformity with rules that we have set. The only restriction on our freedom is that we may under no circumstances contradict the self-established rules. The final criterion of mathematical existence thus becomes freedom from contradiction; that is, mathematical existence can be ascribed to every concept whose use does not enmesh us in contradiction.

Let us now examine a little more critically these three points of view. The realistic position is, as I tried to show in this very place two years ago,[10] untenable since it ascribes to the mind abilities that it does not possess; our thinking consists of tautological transformation, it is incapable of comprehending a reality. Plato assumed a mystical recollection (*anamnesis*) by the soul of a state in which it beheld the ideas face to face, as it were. In any event, this first position is entirely metaphysical and seems wholly unsuitable as the foundation of mathematics. Nonetheless, it is still the source, though often unsuspected, of a great deal of confusion in research on the foundation of mathematics.

With regard to the second, the intuitionistic position, I attempted in last year's lecture series[11] to explain that there is no such thing as pure intuition. To be sure, Kant assigned it a very broad role, but in light of the development of mathematics since his day this view cannot possibly be maintained. Hence the recent supporters of the second position have, in fact, become more modest in their claims. But as to what this allegedly pure intuition can and cannot do, what is consistent with it and what is not – about these matters there is no agreement whatever among the supporters of the intuitionistic position. This is shown very clearly by their answers to the question that concerns us here, "Are there infinite sets and infinite numbers?" Some intuitionists would say that arbitrarily large numbers can perhaps be constructed by pure intuition, but not the set of *all* natural numbers. This group, in other words, would flatly deny the existence of infinite sets. Others of this school hold that while intuition suffices for the construction of the set of all natural numbers, non-denumerably infinite sets are beyond intuition's reach; which is to say they deny the existence of the set of all real numbers. Still others ascribe constructibility, and thereby existence, to certain non-denumerable sets. The intuitionist doctrine is thus seen to rest on very uncertain ground; in glaring contrast to this uncertainty is the gruffness with which the supporters of this position declare meaningless everything that in their opinion is not constructible by pure intuition.

Having rejected the two first positions we must then turn to the third, the logistic interpretation. But before discussing the question "Are there infinite sets and infinite numbers?" from the logistic point of view, it may be useful to point out the difference between the three positions with respect to the axiom of choice, mentioned earlier. A representative of the realistic stand would say: "Whether we are to accept or reject the axiom of choice in logic depends on how reality is constituted; if it is constituted as the axiom of choice asserts, then we must accept it, but if reality is not so constituted, we shall have to reject the axiom. Unfortunately we do not know which is the case, and because of the inadequacy of our means of perception we shall – again unfortunately – never know." An intuitionist would perhaps say: "We must consider whether a set of the kind required by the axiom of choice (that is to say, a set having exactly one member in common with each set of a given system of sets) can be constructed by pure intuition.

For this purpose one would have to select from each set of the given system one member; if the system consisted of infinitely many sets the task would involve infinitely many separate operations; since these cannot possibly be carried out, the axiom of choice is to be rejected." A consistent representative of the logistic position would say: "If the axiom of choice is in truth independent of the other principles of logic, that is to say, if the statement of the axiom of choice as well as the contrary statement is consistent with the other principles of logic, then we are free to accept it or replace it with its contradictory. That is, we can as well operate with a mathematics in which the axiom of choice is taken as a basic principle – a 'Zermelo mathematics' – as with a mathematics in which a contrary axiom is taken as the basis – a 'non-Zermelo mathematics'. The entire question has nothing to do with the nature of reality, as the realists think, or with pure intuition, as the intuitionists think. The question is rather in what sense we decide to use the word 'set'; it is a matter of determining the syntax of the word 'set'."

Let us return to the problem we are mainly concerned with, and consider what answer an adherent of the logistic school would give to the question, "Are there infinite sets and infinite numbers?" He would perhaps reply: "Yes, infinite sets and infinite numbers can be said to exist, provided it is possible to operate with them without contradiction." What is the situation, then, with regard to this freedom from contradiction?

Various philosophers have, in fact, repeatedly raised the objection against Cantor's theory that it would lead to contradictions. The objection might be phrased as follows: "According to Cantor the set of all natural numbers is equivalent to the set of even numbers; this however contradicts the axiom that the whole can never be equal to one of its parts." To refute the objection we must examine the meaning of this alleged axiom. Certainly it cannot mean that reality itself is constituted as the axiom asserts;

for that would be a reversion to the metaphysical realistic position. Its meaning must rather be in a syntactic determination of how we are to use the words 'whole', 'part', 'equal'. We must establish, in other words, that this determination does not correspond completely with the usage of everyday speech or with the language of science. Yet note that even in ordinary linguistic usage it is meaningful to say that the whole may be equal to a part – as, for example, with respect to colour; why, then, should it be forbidden to say that the whole is equal to one of its parts with respect to quantity? The truth is, however, this axiom is used neither in the construction of logic nor of mathematics. Hence no contradiction confronts us here, but merely the fact that as to certain aspects of behaviour infinite sets differ from finite sets. A finite set cannot be equivalent to one of its parts, but an infinite set can be. It can be shown that every infinite set has parts to which it is equivalent, and one can indeed make use of this fact – as Dedekind did – in defining the concept 'infinite set'.

It has been suggested that many other contradictions lurk in the concepts of infinite sets and infinite numbers. These objections, as in the case above, usually consist of showing that certain properties that necessarily belong to finite sets and finite numbers do not belong to infinite sets and infinite numbers. For example, every number must be either even or odd; but Cantor's transfinite cardinal numbers are neither even nor odd; or, every number becomes larger when 1 is added to it: but this is not true of Cantor's transfinites. That transfinite numbers have properties differing from those of finite numbers affords no contradiction; it is not even to be wondered at; it must be so. For if transfinites were in every respect indistinguishable from finite numbers the need for a separate transfinite category would vanish; transfinite numbers could be classed simply as oversized finite numbers. It is the same when we discover a new species of animal – it must differ in some way from the known ones, otherwise it would not be a new species. After disposing of these arguments a single meaningful question remains: "Since they differ so greatly from ordinary numbers is it perhaps not inappropriate to designate transfinites as numbers?" Like many so-called philosophical problems, we are free to consider this one as turning on a simple issue of terminology; though it was for the very purpose of avoiding such purely terminological controversies that Cantor gave his transfinite cardinals the neutral and relatively non-committal name of 'powers'. The controversy, however, intrigues us in much the same way as the once celebrated dispute over whether 'one' is a number, or numbers only begin with 'two'. Let us content ourselves with the indisputable statement that the term 'transfinite numbers' has shown itself wholly suited to its purpose.

Cantor had no difficulty in dealing with objections of this sort; they could not endanger his edifice. But the refutation of a few inadequate proofs of supposed contradictions clearly does not demonstrate that none exist; and serious contradictions have, in fact, been discovered in Cantor's structure. Contradictions have appeared in certain set formations that are sweepingly inclusive, such as the set of all objects, the set of all sets, and the set of all infinite cardinal numbers. Note, however, that the concept of infinity was not the source of these contradictions; instead they arose from certain deficiencies of classical logic as described by Dr. Menger in last year's lecture series.[12] Thus it became evident that what was needed was a reform in logic. This reform consisted mainly in a more careful use of the word 'all', as taught in Russell's theory of logical types.[13] Thereupon the contradictions that had appeared in the theory of sets were successfully explained and eliminated. There is no longer any known contradiction in the present formulation of the theory of sets.

But from the fact that no contradiction is known, it does not necessarily follow that none exists, any more than the fact that in 1900 no okapi was known proved that none existed. This question too was discussed by Dr. Menger in his lecture last year. We face the question, then, "Can any proof be given of freedom from contradiction?" On the basis of present knowledge it may be said that an absolute proof of freedom from contradiction is probably unattainable; every such proof is relative; we can do no more than to relate the freedom from contradiction of one system to that of another. But is not this concession fatal to the logistic position, according to which mathematical existence depends entirely on freedom from contradiction? I think not. For here, as in every sphere of thought, the demand for absolute certainty of knowledge is an exaggerated demand: in no field is such certainty attainable. Even the evidence adduced by many philosophers – the evidence of immediate inner perception exhibited in a statement such as "I now see something white" affords no example of certain knowledge. For even as I formulate and utter the statement "I see something white" it describes a past event and I can never know whether in the period of time that has elapsed, however short, my memory has not deceived me.

There is, then, no absolute proof of freedom from contradiction for the theory of sets and thus no absolute proof of the mathematical existence of infinite sets and infinite numbers. But neither is there any such proof for the arithmetic of finite numbers, nor for the simplest parts of logic. It is a fact, however, that no contradiction is known in the theory of sets, and not a trace of evidence can be found that such a contradiction may turn up. Hence we can ascribe mathematical existence to infinite sets and to Cantor's transfinite numbers with approximately the same certainty as we ascribe existence to finite numbers.

So far we have dealt only with the question whether there are infinite sets and infinite numbers; but no less important, it would appear, is the question whether there are infinite extensions. This is usually phrased in the form: "Is space infinite?" Let us begin by treating this question also from a purely mathematical standpoint.

We must recognize at the outset that mathematics deals with very diverse kinds of space. Here, however, we are interested only in the so-called Riemann spaces discussed by Dr. Nöbeling in his lecture last year,[14] and in particular, in the three-dimensional Riemann spaces. Their exact definition does not concern us; it is sufficient to make the point that such a Riemann space is a set of elements, or points, in which certain subsets, called lines, are the objects of attention. By a process of calculation there can be assigned to every such line a positive number, called the length of the line, and among these lines there are certain ones of which every sufficiently small segment AB is shorter than every other line joining the points A, B. These lines are called the geodesics, or the straight lines of the space in question. Now it may be that in any particular Riemann space there are straight lines of arbitrarily great length; in that case we shall say that this space is of infinite extension. On the other hand, it may also be that in this particular Riemann space the length of all straight lines remains less than a fixed number; then we say that the space is of finite extension. Until the end of the 18th century only a single mathematical space was known and hence it was simply called 'space'. This is the space whose geometry is taught in school and which we call Euclidean space, after the Greek mathematician Euclid who was the first to develop the geometry of this space systematically. And from our definition above, this Euclidean space is of infinite extension.

There are, however, also three-dimensional Riemann spaces of finite extension; the best known of these are the so-called spherical spaces (and the closely related elliptical ones), which are three-dimensional analogues of a spherical surface. The surface of a sphere can be conceived as two-dimensional Riemann space, whose geodesics, or "straight" lines, are arcs of great circles. (A great circle is a circle cut on the surface of a sphere by a plane passing through the centre of the sphere, as for instance, the equator and the meridians of longitude on the earth.) If r is the radius of the sphere, then the full circumference of a great circle is $2\pi r$; that is to say, no great circle can be longer than $2\pi r$. Hence the sphere considered as two-dimensional Riemann space is a space of finite extension. With regard to three-dimensional spherical space the situation is fully analogous; this also is a space of finite extension. Nevertheless, it has no boundaries; it is unbounded as the surface of a sphere is unbounded; one can keep walking along one of its straight lines without ever being stopped by a boundary of the space. After a finite time one simply comes back to the starting point, exactly as if one had kept moving farther and farther along a great circle of a spherical surface. In other words, we can make a circular tour of spherical space, just as easily as we can make a circular tour of the earth.

Thus we see that in a mathematical sense there are spaces of infinite extension (e.g., Euclidean space) and spaces of finite extension (e.g., spherical and elliptical spaces). Yet, this is not at all what most persons have in mind when they ask "Is space infinite?" They are asking, rather, "Is the space in which our experience and in which physical events take place of finite or of infinite extension?"

So long as no mathematical space other than Euclidean space was known, everyone naturally believed that the space of the physical world was Euclidean space, infinitely extended. Kant, who explicitly formulated this view, held that the arrangement of our observations in Euclidean space was an intuitional necessity; the basic postulates of Euclidean geometry are synthetic, *a priori* judgments.

But when it was discovered that in a purely mathematical sense spaces other than Euclidean also 'existed', (that is, led to no logical contradictions) it became possible to question the view that the space of the physical world must be Euclidean space. And the idea developed that it was a question of experience, that is, a question that must be decided by experiment, whether the space of the physical world was Euclidean or not. Gauss actually made such experiments. But after the work of Henri Poincaré, the great mathematician of the end of the 19th century,[15] we know that the question expressed in this way has no meaning. To a considerable extent we have a free choice of the kind of mathematical space in which we arrange our observations. The question does not acquire meaning until it is decided how this arrangement is to be carried out. For the important thing about Riemann space is the manner in which each of its lines is assigned a length, that is, how lengths are measured in it. If we decide that measurements of length in the space of physical events shall be made in the way they have been made from earliest times, that is, by the application of 'rigid' measuring rods, then there is meaning in the question whether the space of physical events, considered as a Riemann space, is Euclidean or non-Euclidean. And the same holds for the question whether it is of finite or of infinite extension.

The answer that many perhaps are prompted to give, "Of course, by this method of measurement physical space becomes a mathematical space of infinite extension," would be somewhat too hasty. As background for a brief discussion of this problem we must first give a short and very simple statement of certain mathematical facts. Euclidean space is characterized by the fact that the sum of the three angles of a triangle in such space is 180 degrees. In spherical space the sum of the angles of every triangle is greater than 180 degrees, and the excess over 180 degrees is greater the larger the triangle in relation to the sphere. In the two-dimensional analogue of spherical space, the surface of a sphere, this point is presented to us very clearly. On the surface of a sphere, as already mentioned, the counterpart of the straight-line triangle of spherical space is a triangle whose sides are arcs of great circles, and it is a well-known proposition of elementary geometry that the sum of the angles of a spherical triangle is greater than 180 degrees, and that the excess over 180 degrees is greater the larger the surface area of the triangle. If a further comparison be made of spherical triangles of equal area on spheres of different sizes, it may be seen at once that the excess of the sum of the angles over 180 degrees is greater the smaller the diameter of the sphere, which is to say, the greater the curvature of the sphere. This gave rise to the adoption of the following terminology (and here it is simply a matter of terminology, behind which nothing whatever secret is hidden): A mathematical space is called 'curved' if there are triangles in it the sum of whose angles deviates from 180 degrees. It is 'positively curved' if the sum of the angles of every triangle in it (as in elliptical and spherical spaces) is greater than 180 degrees, and 'negatively curved' if the sum is less than 180 degrees – as is the case in the 'hyperbolic' spaces discovered by Bolyai and Lobachevsky.

From the mathematical formulations of Einstein's General Theory of Relativity it now follows that, if the previously mentioned method of measurement is used as a basis, space in the vicinity of gravitating masses must be curved in a 'gravitational field'.[16] The only gravitational field immediately accessible, that of the earth, is much too weak for us to be able to test this assertion directly. It has been possible, however, to prove it indirectly by the deflection of light rays – as determined during total eclipses – in the much stronger gravitational field of the sun. So far as our present experience goes, we can say that if, by using the measuring methods mentioned above, we turn the space of physical events into a mathematical Riemann space, this mathematical space will be curved, and its curvature will, in fact, vary from place to place, being greater in the vicinity of gravitating masses and smaller far from them.

To return to the question that concerns us: Can we now say whether this space will be of finite or infinite extension? What has been said so far is not sufficient to give the answer; it is still necessary to make certain rather plausible assumptions. One such assumption is that matter is more or less evenly distributed throughout the entire space of the universe: that is to say, in the universe as a whole there is a spatially uniform density of mass. The observations of astronomers to date can, at least with the help of a little good will, be brought into harmony with this assumption. Of course it can be true only when taken in the sense of a rough average, in somewhat the same sense as it can be said that a piece of ice has on the whole the same density throughout. Just as the mass of the ice is concentrated in a great many very small particles, separated by intervening spaces that are enormous in relation to the size of these particles, so the stars in world-space are separated by intervening spaces that are enormous in relation to the size of the stars. Let us make another quite plausible assumption, namely, that the universe, taken by and large, is stationary in the sense that this average constant density of mass remains unchanged. We consider a piece of ice stationary, even though we know that the particles that constitute it are in active motion; we may likewise deem the universe to be stationary, even though we know the stars to be in active motion. With these assumptions, then, it follows from the principles of the General Theory of Relativity that the mathematical space in which we are to interpret physical events must on the whole have the same curvature throughout. Such a space, however, like the surface of a sphere in two dimensions, is ncessarily of finite extension. In other words, if we use as a basis the usual method of measuring length and wish to arrange physical events in a mathematical space, and if we make the two plausible assumptions mentioned above, the conclusion follows that this space must be of finite extension.

I said that the first of our assumptions, that of the equal density of mass throughout space, conforms somewhat with observations. Is this also true of the second assumption, as to the constant density of mass with respect to time? Until recently this opinion was tenable. Now, however, certain astronomical observations seem to indicate – again speaking in broad terms – that all heavenly bodies (fixed stars and nebulae) are moving away from us with a velocity that increases the greater their distance from us, the velocity of those that are the farthest away from us, but still within reach of study, being quite fantastic. But if this is so, the average density of mass of the universe cannot possibly be constant in time; instead it must continually become smaller. Then if the remaining features of our picture of the universe are maintained, it would mean that we must assume that the mathematical space in which we interpret physical events is variable in time. At every instant it would be a space with (on the average) a constant positive curvature, that is to say, of finite extension, but the curvature would be continually decreasing while the extension would be continually increasing. This interpretation of physical events occurring in an expanding space turns out to be entirely workable and in accord with the General Theory of Relativity.

But is this the only theory consistent with our experience to date? I said before that the assumption that the space of the universe was on the whole of uniform density could fairly well be brought into harmony with astronomical observations. At the same time these observations do not contradict the entirely different assumption that we and our system of fixed stars are situated in a region of space where there is a strong concentration of mass, while at increasing distances from this region the distribution of mass keeps getting sparser. This would lead us – still using the ordinary method of measuring length – to conceive of the physical world as situated in a space that has a certain curvature in the neighbourhood of our fixed star system, a curvature, however, that grows smaller and smaller further away.[17] Such a space can of course be of infinite extension. Similarly the phenomenon that the stars are in general receding from us, with greater velocity the farther away they are, can be quite simply explained as follow:[18] Assume that at some time many masses with completely different velocities were concentrated in a relatively small region of space, let us say in a sphere K. In the course of time these masses will then, each with its own particular velocity, move out of this region of the space. After a sufficient time has elapsed, those that have the greatest velocities will have moved farthest away from the sphere K, those with lesser velocities will be nearer to K, and those with the lowest velocities will still be very close to K or even within K. Then an observer within K, or at least not too far removed from K, will see the very picture of the stellar world that we have described above. The masses will on the whole be moving away from him, and those farthest away will be moving with the greatest velocities. We would thus have an interpretation of the physical world in an entirely different kind of mathematical space – that is to say, in an infinitely extended space.

In summary we might very well say that the question: "Is the space of our physical world of infinite or of finite extension?" has no meaning as it stands. It does not become meaningful until we decide how we are to go about getting the observed events of the physical world into a mathematical space, that is, what assumptions must be made and what logical requirements must be satisfied. And this in turn leads to the question: "Is a finite or an infinite mathematical space better adapted for the arrangement and interpretation of physical events?" At the present stage of our knowledge we cannot give any reasonably well-founded answer to this question. It appears that mathematical spaces of finite and of infinite extension are almost equally well suited for the interpretation of the observational data thus far accumulated.

Perhaps at this point confirmed 'finitists' will say: "If this is so, we prefer the scheme based on a space of finite extension, since any theory incorporating the concept of infinity is wholly unacceptable to us." They are free to take this view if they wish, but they must not imagine thereby to have altogether rid themselves of infinity. For even the finitely extended Riemann spaces contain infintely many points, and the mathematical treatment of time is such that each time-interval, however small, contains infinitely many time-points.

Must this necessarily be so? Are we in truth compelled to lay the scene of our experience in a mathematical space or in a mathematical time that consists of infinitely many points? I say no. In principle one might very well conceive of a physics in which there were only a finite number of space points and a finite number of time points – in the language of the theory of relativity, a finite number of 'world points'. In my opinion neither logic nor intuition nor experience can ever prove the impossibility of such a truly finite system of physics. It may be that the various theories of the atomic structure of matter, or today's quantum physics, are the first foreshadowings of a future finite physics. If it ever comes, then we shall have returned after a prodigious circular journey to one of the starting points of western thought, that is, to the Pythagorean doctrine that everything in the world is governed by the natural numbers. If the famous theorem of the right-angle triangle rightly bears the name of Pythagoras then it was Pythagoras himself who shook the foundations of his doctrine that everything was governed by the natural numbers. For from the theorem of the right triangle there follows the existence of line segments that are incommensurable, that is, whose relationship with one another cannot be expressed by the natural numbers. And since no distinction was made between mathematical existence and physical existence, a finite physics appeared impossible. But if we are clear on the point that mathematical existence and physical existence mean basically different things; that physical existence can never follow from mathematical existence; that physical existence can in the last analysis only be proved by observation; and that the mathematical difference between rational and irrational forever transcends any possibility of observation – then we shall scarcely be able to deny the possibility in principle of a finite physics. Be that as it may, whether the future produces a finite physics or not, there will remain unimpaired the possibility and the grand beauty of a logic and a mathematics of the infinite.

NOTES

1 Cf. P. Duhem, *Études sur Léonard de Vinci, ceux qu'il a lus et ceux qui l'ont lu,* Seconde série (Paris, 1909), IX Léonard de Vinci et les deux infinis.

2 [*Principia* I.xxvi].

3 [Letter XIX to Foucher, *Philosophische Schriften* (ed. Gerhardt) Vol. I, p. 416. This passage is quoted on the title-page of Bolzano, see note 5 below.]

4 [In a letter to Schumacher quoted by Fraenkel, *Einleitung*, p. 1 (see note 6 below). Letter of 12 July 1831, *Werke* Vol. 8, p. 216.]

5 *Die Paradoxien des Unendlichen* (Leipzig, 1950), new edition with notes by H. Hahn (Leipzig, 1920) [E. T. *Paradoxes of the Infinite* (London, 1950)].

6 G. Cantor's fundamental articles are contained in Georg Cantor, *Gesammelte Abhandlungen*, ed. E. Zermelo (Berlin, 1932). This volume also contains a portrait of Cantor and a biography composed by A. Fraenkel. For further study of set theory A. Fraenkel's *Einleitung in die Mengenlehre*[3] (Berlin, 1928) is to be recommended. [E. T. of some of Cantor's articles in G. Cantor, *Contributions to the Founding of the Theory of Transfinite Numbers*, tr. P. E. B. Jourdain (London & Chicago, 1915); E. T. of Fraenkel as *Foundations of Set Theory* (Amsterdam, 1958).]

7 For sets, *M*, consisting of one or two members, the rule holds only if one adds to the proper partial sets the empty set (null-set) possessing no members and the set *M* itself.

8 For further reading on the continuum problem see W. Sierpiński, *Hypothèse du continu* (Monografje Matematyczne, Tom. IV (Warsaw & Lvov, 1934). [The problems discussed here and in the next note have been decisively advanced by K. Gödel and P. J. Cohen. The former showed, "The axiom of choice and Cantor's generalized continuum hypothesis . . . are consistent with the other axioms of set theory if these axioms are consistent" (*The Consistency of the Axiom of Choice* &c, Princeton, N. J., 1940, Introduction). The latter showed "that CH [the continuum hypothesis] cannot be proved from ZF [Zermelo-Fraenkel set theory] (with AC [the axiom of choice] included), and that AC cannot be proved from ZF" (*Set Theory and the Continuum Hypothesis*, New York, 1966 Introduction to Chapter IV)].

9 For further reading on the axiom of choice see W. Sierpiński, L'axiome de M. Zermelo et son rôle dans la théorie des ensembles et l'analyse. Bull. de l'Acad. des Sciences de Cracovie, Classes des sciences math. et nat., Série A,

1918, pp. 97–152; W. Sierpiński, Leçons sur les nombres transfinis (Paris, Gauthier-Villars, 1928). Chap. VI. [See also note 8 above.]

10 In the series of lectures in aid of a memorial to Ludwig Boltzmann. This lecture was published as H. Hahn, *Logik, Mathematik und Naturerkennen* (Einheitswissenschaft, No. 2, Vienna 1933). [It will be published in E. T. with the other *Einheitswissenschaft* pamphlets in a future volume of this Collection.]

11 *Krise und Neuaufbau in den exakten Wissenschaften* (Leipzig & Vienna 1933): H. Hahn, "Die Krise der Anschauung" [E. T. Chapter VIII of the present volume.]

12 *Ibid.*: K. Menger, 'Die neue Logik' [E. T. Chapter I of his *Selected Papers in Logic and Foundations, Didactics, Economics*, (Dordrecht, Boston, and London, 1979) in this Collection.]

13 Cf. A. N. Whitehead and B. Russell, *Principia Mathematica*[2] (Cambridge, 1925) Vol. I pp. 37ff.

14 *Krise und Neuaufbau in den exakten Wissenschaften*: G. Nöbeling, 'Die vierte Dimension und der krumme Raum'.

15 H. Poincaré, *La Science et l'Hypothèse* (Paris, 1902) [E. T. *Science and Hypothesis* (London, 1905)].

16 For what follows see the easily intelligible presentation of A. Haas, *Kosmologische Probleme der Physik* (Leipzig, 1934) and the detailed but much more difficult one by H. Weyl, *Raum, Zeit, Materie*[5] (Berlin, 1923) [E. T. *Space, Time, Matter* (London, 1922)].

17 We are thus led to mathematical spaces that are three-dimensional analogues of a paraboloid of rotation. Cf. H. Weyl, op. cit. p. 257.

18 The following interpretation is due to E. A. Milne. Cf. A. Haas, op. cit., p. 59 and E. Freundlich, *Die Naturwissenschaften* **21** (1933) p. 54.

BIBLIOGRAPHY OF BOOKS AND ARTICLES BY HANS HAHN

What follows is essentially the bibliography compiled by K. Mayrhofer shortly after Hahn's death and printed in *Monatshefte für Mathematik und Physik*, Volume 41 pp. 231–8. Reviews are not included. They generally had no philosophical content; indeed Chapter VI above was the only exception we could find.

1. BOOKS AND MONOGRAPHS

Weiterentwicklung der Variationsrechnung in den letzten Jahren. With E. Zermelo. Enzyklopädie d. mathem. Wissensch. (B. G. Teubner, Leipzig.) II A, 8a (1904).

Arithmetik, Mengenlehre, Grundbegriffe der Funktionenlehre. E. Pascal, Repertorium d. höheren Mathem. (B. G. Teubner, Leipzig.) Vol. I, 1, Ch. 1 (1910).

Variationsrechnung. Ibidem. Vol. I, 2, Ch. XIV (1927).

Die Theorie der Integralgleichungen und Funktionen unendlich vieler Variabeln und ihre Anwendung auf die Randwertaufgaben bei gewöhnlichen und partiellen Differentialgleichungen. With L. Lichtenstein and J. Lense.) Ibidem. Vol. I, 3, Ch. XXIV (1929).

B. Bolzano, Paradoxien des Unendlichen. Notes by H. Hahn. (F. Meiner, Leipzig 1920.)

Theorie der reellen Funktionen I. (J. Springer, Berlin 1921.)

Einführung in die Elemente der höheren Mathematik. With H. Tietze. (S. Hirzel, Leipzig 1925.)

Reelle Funktionen I. (Akad. Verlagsges., Leipzig 1932.)

Überflüssige Wesenheiten (*Occams Rasiermesser*). (A. Wolf, Vienna 1930.) [E. T. Chapter I above].

Logik, Mathematik und Naturerkennen. Einheitswissenschaft, Heft 2. (Gerold & Co., Vienna 1933.)

Krise und Neuaufbau in den exakten Wissenschaften. Fünf Wiener Vorträge. H. Hahn: Die Krise der Anschauung. (F. Deuticke, Leipzig and Vienna 1933.) [E. T. Chapter VII above].

Alte Probleme – Neue Lösungen in den exakten Wissenschaften. Fünf Wiener Vorträge. Includes H. Hahn: Gibt es Unendliches? (F. Deuticke, Leipzig and Vienna 1934.) [E. T. Chapter VIII above].

Set Functions. Continuation of *Reelle Funktionen*, completed by A. Rosenthal (University of New Mexico Press, Albuquerque, N. M., 1948.)

2. PUBLICATIONS IN PERIODICALS

Acta mathematica: Über eine Verallgemeinerung der Fourierschen Integralformel. **49** (1926).

Annali di Matematica: Über die Abildung einer Strecke auf ein Quadrat. (III) **21** (1913).

Annali di Pisa: Über die Multiplikation total-additiver Mengenfunktionen. (II) **2** (1933).

Anzeiger der Akademie der Wissenschaften in Wien: Über die nichtarchimedischen Gröβensysteme. **44** (1907). – Über Extremalenbögen, deren Endpunkt zum Anfangspunkt konjugiert ist. **46** (1907). – Über einfach geordnete Mengen. **50** (1913). – Über die Darstellung gegebener Funktionen durch singuläre Integrale. **53** (1916). – Über halbstetige und unstetige Funktionen. **54** (1917). – Finige Anwendungen der Theorie der singulären Integrale **55** (1918). – Über irreduzible Kontinua. **58** (1921). – Dankschreiben f. d. Verleihung d. R. Lieben-Preises. **58** (1921). – Dankschreiben f. seine Wahl zum korr. Mitglied. **58** (1921). – Über ein Existenztheorem der Variationsrechnung. **62** (1925). – Über die Methode der arithmetischen Mittel **62** (1925). – Über additive Mengenfunktionen. **65** (1928). – Über stetige Streckenbilder. **65** (1928). – Über unendliche Reihen und total-additive Mengenfunktionen. **65** (1928). – Über den Integralbegriff. **66** (1929). – Über separable Mengen. **70** (1933).

Archiv der Mathematik und Physik: Über die Menge der Konvergenzpunkte einer Funktionenfolge. III, **28** (1919).

Atti del Congresso Internazionale dei Matematici, Bologna 1928: Über stetige Streckenbilder.

Bulletin of the Calcutta Mathematical Society: Über unendliche Reihen und absolutadditive Mengenfunktionen. **20** (1930).

Denkschriften der Akademie der Wissenschaften in Wien, math-naturw. Kl.: Über die Darstellung gegebener Funktionen durch singuläre Integrale I und II. **93** (1916).

Erkenntnis: Die Bedeutung der wissenschaftlichen Weltauffassung, insbesondere für Mathematik und Physik. **1** (1930) [E. T. Chapter II above]. –

Diskussion zur Grundlegung der Mathematik. **2** (1931) [E. T. Chapter III above].

Festschrift der 57. Versammlung Deutscher Philologen und Schulmänner in Salzburg: Über den Integralbegriff. (1929).

Forschungen und Fortschritte: Empirismus, Mathematik, Logik. **5** (1929) [E. T. Chapter IV above].

Fundamenta mathematica: Über die Komponenten offener Mengen. **2** (1921).

Jahresbericht der Deutschen Mathematiker – Vereinigung. Bericht über die Theorie der linearen Integralgleichungen. **20** (1911). – Über die allgemeinste ebene Punktmenge, die stetiges Bild einer Strecke ist. **23** (1914). – Über Fejérs Summierung der Fourierschen Reihe. **25** (1916). – Über stetige Funktionen ohne Ableitung. **26** (1918). – Über die Vertauschbarkeit der Differentiationsfolge. **27** (1919). – Arithmetische Bemerkungen. (Entgegnung auf Bemerkungen des Herrn J. A. Gmeiner.) **30** (1921). – Schlußbemerkungen hiezu. **30** (1921). – Über Funktionaloperationen. **31** (1922). – Über die Darstellung willkürlicher Funktionen durch bestimmte Integrale. (Bericht.) **30** (1921). – Über Fouriersche Reihen und Integrale. **33** (1924).

Journal für die reine und angewandte Mathematik: Über lineare Gleichungssysteme in linearen Räumen. **157** (1927).

Mathematische Annalen: Bemerkungen zur Variationsrechnung. **58** (1904). – Über die Herleitung der Differentialgleichungen der Variationsrechnung. **63** (1907). – Über räumliche Variationsprobleme. **70** (1911).

Mathematische Zeitschrift: Über das Interpolationsproblem. **1** (1918). – Über Funktionen mehrerer Veränderlicher, die nach jeder einzelnen Veränderlichen stetig sind. **4** (1919). – Über die stetigen Kurven der Ebene. **9** (1921).

Monatshefte für Mathematik und Physik: Zur Theorie der zweiten Variation einfacher Integrale. **14** (1903). – Über die Lagrangesche Multiplikatorenmethode in der Variationsrechnung. **14** (1903). – Über Funktionen zweier komplexer Veränderlichen. **16** (1905). – Über den Fundamentalsatz der Integralrechnung. **16** (1905). – Über punktweise unstetige Funktionen. **16** (1905). – Über einen Satz von Osgood in der Variationsrechnung. **17** (1906). – Über das allgemeine Problem der Variationsrechnung. **17** (1906). – Bemerkungen zu den Untersuchungen des Herrn M. Fréchet: Sur quelques points du calcul fonctionnel. **19** (1908). – Über die Anordnungssätze der Geometrie. **19** (1908). – Über Bolzas fünfte notwendige Bedingung in der Variationsrechnung. **20** (1909). – Über Variationsprobleme mit variablen Endpunkten. **22** (1911). – Über die Integrale des Herrn Hellinger und die Orthogonalinvarianten der quadratischen Formen von unendlich vielen Veränderlichen. **23** (1912). – Ergänzende Bemerkung zu meiner Arbeit über den Osgoodschen Satz in Band 17 dieser Zeitschrift. **24** (1913).

– Über eine Verallgemeinerung der Riemannschen Integraldefinition. **26** (1915). – Über Folgen linearer Operationen. **32** (1922). – Über Reihen mit monoton abnehmenden Gliedern. **33** (1923). – Die Äquivalenz der Cesàroschen und Hölderschen Mittel. **33** (1923).

Die Naturwissenschaften: Mengentheoretische Geometrie. **17** (1929).

Rendiconti del Circolo Matematico di Palermo: Über den Zusammenhang zwischen den Theorien der zweiten Variation und der Weierstraβschen Theorie der Variationsrechnung. **29** (1910).

Sitzungsberichte der Akademie der Wissenschaften in Wien, math.-naturw. Klasse: Über die nicht-archimedischen Gröβensysteme. **116** (1907). – Über Extremalenbogen, deren Endpunkt zum Anfangspunkt konjugiert ist. **118** (1909). – Über einfach geordnete Mengen. **122** (1913). – Über Annäherung an Lebesguesche Integrale durch Riemannsche Summen. **123** (1914). – Mengentheoretische Charakterisierung der stetigen Kurve. **123** (1914). – Über halbstetige und unstetige Funktionen. **126** (1917). – Einige Anwendungen der Theorie der singulären Integrale. **127** (1918). – Über irreduzible Kontinua. **130** (1921). – Über die Lagrangesche Multiplikatorenmethode. **131** (1922). – Über ein Existenztheorem der Variationsrechnung. **134** (1925). – Über die Methode der arithmetischen Mittel in der Theorie der verallgemeinerten Fourierschen Integrale. **134** (1925).

H. Weber-Festschrift, 1922: Allgemeiner Beweis des Osgoodschen Satzes der Variationsrechnung für einfache Integrale.

Zeitschrift für Mathematik und Physik: Über das Strömen des Wassers in Röhren und Kanälen (with G. Herglotz und K. Schwarzschild). **51** (1904).

INDEX

~ indicates repetition of the catchword or of that part of it preceding /.

Ampère, A.-M. 82
analytic, so-called ~ judgements, 22
a posteriori, 73, *see* experience
a priori, 21, 26, 73
Aquinas, 103
arbitrariness, 62–5, 71
Aristotle, 4, 103
arithmetic, 55, 76
  foundations of ~, 58
arithmetization of analysis, 66–7
axiom, 35–6, 72, *see* proposition
  ~ atization, 13, 26–7, 54
  ~ of choice, 34–5, 113
  ~ of infinity, 35
  ~ of reducibility, 25

Being, true ~, 1–3, 28, 43–6, *see* ideas
Bolzano, B., xi, 77, 93, 104, 130
Brouwer, L.E.J., 25–6, 32, 90, 102

Cantor, G., 53, 71, 104–31
Carnap, R., x–xiv, 34, 93
Cauchy, A., 93, 104
class, 17, 18, 71, *see* set
  ~ of bodies, 13–14
  ~ of colours &c, 17
  elimination of ~ es, 37
  ~ of experiences, 11
  ~ of sets, 15, 37
Cohen, P.J., 130
constituting a term, 22, 27–8
constructibility, 36–7
continuum, power of the ~, 110
contradictions / in set theory, 53, 71, 121
  freedom from ~, 36, 121

Dedekind, R., 93, 119
Democritus, 4
Descartes, R., 103–4
dimensionality, 94
dualism, 20–1

Einstein's theory of relativity, 74 125–7
Eleatic philosophers, 3–4, 43, 45
elementary methods, 68–9
empiricism, xiv, 20–1, 31, 33, 39–42, 76
entities, other-worldly ~, 7, 11–2, 14, 16, 18, *see* being, ideas, Occam's razor
epistemology, 7
existence / in analysis, 36

~ in mathematics, 114, 116, 121, 123, 129
experience, 10–2, 15–8, 21, 26, 28, 31, 39, 41, 73, 75

Finite) (infinite, 57, 59, 70, 105–30
formalism, 25–6, 32
Fraenkel, A., 130
Frege, A., 15

Gauss, C.F., 104, 124
geometry, 12, 66, 74
~ and analysis, 66, 68
non-Archimedean ~, 99–102
non-Euclidean ~, 99–101, 112–3
~ and intuition, 76–101
logicization, axiomatization of ~, 26–7, 35, 67
Gödel, K., xiii, xviii, 130

Heisenberg, W., 101
Heyting, A., 32
Hilbert, D., 25–6, 32, 35–6
Hume, D., xi, 4

Idealism, 4
ideas, world of ~, 3–5, 14, *see* Plato
identity, absolute ~, 25
induction, complete, 61
infinite, 103–131, *see* finite
denumerably) (non-denumerably ~, 106, 107, 109, 111, 117
intuition, 55, 73
geometry and ~, 27, 55, 76–102
mystical ~, 20, 30, 31
primitive ~ of the number series, 32
pure ~, 31, 76
intuitionism, 25–6, 32, 34, 36, 115–6

Kant, I., xi, 10, 26, 43, 73–6, 101, 115–6

Language, xii, 9, 28–9
structure of ~) (structure of world 8, 16
Leibniz, G.W.F., xi, 43, 77, 78, 104, 114
Locke, J., 4, 104
Loewy, A., 70
logic, 21–3, 32, 34, 40–2, 50, 53–4, 93 *see* mathematics, tautological
~ al analysis, 84–5, 87
~ ism, 34, 116, 118
~ ization of geometry, arithmetic, analysis, 26–7, 35, 67
symbolic ~, 9, 29, 71

Mach, E., xi, 17
mathematics, 9, 14, 20–1, 23, 25, 32, 39–40, 42, 76, 93 *see* tautological foundations of ~, 31, 34, 52, 58
~ and logic, 52–3, 55–6, 93, 116
Menger, K., ix–xviii, 76, 94, 101–2, 120, 131
metaphysics, xiv, 9, 14, 22, 28, 30, 76, 119

Neumann, J. von, 32
Newton, I., 77–78
Nöbeling, G., 122
nominalism, 5, 6, 103, 116
numbers, 14, 15, 17, 28
algebraic ~, 108, 110

cardinal ~, 36, 58, 60, 106
extension of system of ~, 64
natural ~, 37, 54, 56–7, 59, 63–4, 70, 106, 117
ordinal ~, 58, 60
rational ~, 63, 71, 75
real, irrational ~, 71, 75, 99, 109–10
transcendental ~, 110
transfinite cardinal ~, 106, 111–3, 119–21
transfinite ordinal ~, 71

Occam's razor, 4–6, 9, 12, 14–5, 17–9

Parapsychology, xv–xvii
Peano, G., 53, 55, 62, 70, 85
permanence, principle of ~, 61
philosophy, world-affirming) (world-denying ~, 1–19
physics, 12, 20, 22, 24, 27, 41, 75, 100, 129
Planck, M., 43
Plato, 3–4, 14, 23, 43–4, 114, 116
Poincaré, H., xii, 124, 131
Pringsheim, A., 51–72
projection, isomorphic ~, 58, 60
one-to-one ~, 60
proposition, fundamental unproved ~, 57, 72, *see* axiom
pseudo-propositions, 20, 28

Ramsey, F.P., 34
rationalism, xiv, 4, 20–1
realism, 5–6, 34, 36, 115–7, 119
reality, 'in ~', 46–50
Reimann spaces, 122–5, 128–9
rigour, absolute) (greatest possible ~, 52–4
Russell, B., xi–xii, 4, 11–3, 15, 17, 24–5, 27, 34–5, 37, 42, 70–1, 76, 101, 113, 120, 131

Schubert, H., 62
senses, deception by the ~, 2, 44, 46–7
world of the ~, 1, 3, 7, 17–9, 28, 43, *see* entities
set, 56, 103–131, *see* class
~ of distinguishable things, 57
equivalence of ~ s, 105
~ -theory, 53–4, *see* contradictions
Sierpiński, W., 96, 130
simultaneity, 10–11
space, 9, 12, 14, 17–8, 28, 67, 73, 75
~ of geometry a logical construct, 98
Study, E., 67
subject-predicate structure, 8, 16, 28
substances and properties 16, 28

Tangent problem, 78
tautological, ~ character of logic, mathematics, 7, 23–5, 34
~ transformation, 21–2, 24, 28, 33, 46
thought, 8, 20
~) (experience, 21, 28, 31
~) (intuition, 84
~ parallel to the world, 24, 41
~) (sense, 2, 7, 44, 45
time, 9–10, 14, 17–8, 74–5, *see* simultaneity
types, theory of ~, 25, 120

Universals, 5, 18
Urysohn, P., 94

Waismann, F., x–xiv, 37

Weierstrass, C., 77, 82, 93
Whitehead, A.N., 25, 27, 70, 101, 131, *see* Russell
Wittgenstein, L., xii, 24–5, 34, 37
world, *see* entities, philosophy, senses

Zermelo, E., 112, 118, 130

VIENNA CIRCLE COLLECTION

1. OTTO NEURATH, *Empiricism and Sociology.* Edited by Marie Neurath and Robert S. Cohen. With a Section of Biographical and Autobiographical Sketches. Translations by Paul Foulkes and Marie Neurath. 1973, xvi + 473 pp., with illustrations. ISBN 90-277-0258-6 (cloth), ISBN 90-277-0259-4 (paper).

2. JOSEF SCHÄCHTER, *Prolegomena to a Critical Grammar.* With a Foreword by J. F. Staal and the Introduction to the original German edition by M. Schlick. Translated by Paul Foulkes. 1973, xxi + 161 pp. ISBN 90-277-0296-9 (cloth), ISBN 90-277-0301-9 (paper).

3. ERNST MACH, *Knowledge and Error. Sketches on the Psychology of Enquiry.* Translated by Paul Foulkes. 1976, xxxviii + 393 pp. ISBN 90-277-0281-0 (cloth), ISBN 90-277-0282-9 (paper).

4. MARIA REICHENBACH and ROBERT S. COHEN, *Hans Reichenbach: Selected Writings, 1909-1953 (Volume One).* 1978, in press. ISBN 90-277-0291-8 (cloth), ISBN 90-277-0292-6 (paper). *Hans Reichenbach: Selected Writings, 1909-1953 (Volume Two).* 1978, in press. ISBN 90-277-0909-2 (cloth), ISBN 90-277-0910-6 (paper). Sets: ISBN 90-277-0892-4 (cloth), ISBN 90-277-0893-2 (paper).

5. LUDWIG BOLTZMANN, *Theoretical Physics and Philosophical Problems. Selected Writings.* With a Foreword by S. R. de Groot. Edited by Brian McGuinness. Translated by Paul Foulkes. 1974, xvi + 280 pp. ISBN 90-277-0249-7 (cloth), ISBN 90-277-0250-0 (paper).

6. KARL MENGER, *Morality, Decision, and Social Organization. Toward a Logic of Ethics.* With a Postscript to the English Edition by the Author. Based on a translation by E. van der Schalie. 1974, xvi + 115 pp. ISBN 90-277-0318-3 (cloth), ISBN 90-277-0319-1 (paper).

7. BÉLA JUHOS, *Selected Papers on Epistemology and Physics.* Edited and with an Introduction by Gerhard Frey. Translated by Paul Foulkes. 1976, xxi + 350 pp. ISBN 90-277-0686-7 (cloth), ISBN 90-277-0687-5 (paper).

8. FRIEDRICH WAISMANN, *Philosophical Papers.* Edited by Brian McGuinness with an Introduction by Anthony Quinton. Translated by Hans Kaal (Chapters I, II, III, V, VI and VIII and by Arnold Burms and Philippe van Parys. 1977, xxii + 190 pp. ISBN 90-277-0712-X (cloth), ISBN 90-277-0713-8 (paper).

9. FELIX KAUFMANN, *The Infinite in Mathematics. Logico-mathematical writings.* Edited by Brian McGuinness, with an Introduction by Ernest Nagel. Translated from the German by Paul Foulkes. 1978, xviii + 236 pp. ISBN 90-277-0847-9 (cloth), ISBN 90-277-0848-7 (paper).

10. KARL MENGER, *Selected Papers in Logic and Foundations, Didactics, Economics.* 1978, in press. ISBN 90-277-0320-5 (cloth), ISBN 90-277-0321-3 (paper).

11. HENK L. MULDER and BARBARA F. B. VAN DE VELDE-SCHLICK, *Moritz Schlick: Philosophical Papers Volume I (1909-1922).* Translated by Peter Heath. 1978, xxxviii + 376 pp. ISBN 90-277-0314-0 (cloth), ISBN 90-277-0315-9 (paper).

12. EINO SAKARI KAILA, *Reality and Experience. Four Philosophical Essays.* Edited by Robert S. Cohen. 1978, in press. ISBN 90-277-0915-7 (cloth), ISBN 90-277-0919-X (paper).

13. HANS HAHN, *Empiricism, Logic, and Mathematics, Philosophical Papers.* Edited by Brian McGuinness. 1980. ISBN 90-277-1056-1 (cloth), ISBN 90-277-1066-X (paper).